Advanced Studies in Theoretical and Applied Econometrics

AF822777

Volume 51

More information about this series at http://www.springer.com/series/5667

Daniel A. Griffith • Jean H. P. Paelinck

Morphisms for Quantitative Spatial Analysis

Springer

Daniel A. Griffith
University of Texas at Dallas
Richardson, Texas
USA

Jean H. P. Paelinck
George Mason University
Fairfax, VA, USA

ISSN 1570-5811 ISSN 2214-7977 (electronic)
Advanced Studies in Theoretical and Applied Econometrics
ISBN 978-3-030-10224-1 ISBN 978-3-319-72553-6 (eBook)
https://doi.org/10.1007/978-3-319-72553-6

Printed on acid-free paper

This Springer imprint is published by Springer Nature
The registered company is Springer International Publishing AG
The registered company address is: Gewerbestrasse 11, 6330 Cham, Switzerland

To Juliette's memory

Preamble

In our previous book (Daniel A. Griffith and Jean H. P. Paelinck, *Non-standard Spatial Statistics and Spatial Econometrics*, Springer, Berlin Heidelberg, 2011), we introduce a certain number of novel analytical approaches to the two research fields mentioned. This book is a follow-up to the one just mentioned, in that it presents a certain number of formal approaches borrowed from other fields of investigation and applies them to empirical materials. As such, it tries to demonstrate the fruitfulness of methodological multidisciplinarity in discovering spatial relations in geography and economics. And as we said previously: *lectorem saluti* …

Richardson, TX — Daniel A. Griffith
Fairfax, VA, USA — Jean H. P. Paelinck

Contents

Part I
Spatial Statistics

Chapter 1
Introduction to Part I: Spatial Statistics

1.1 Introduction

This book treats the notion of morphisms in spatial analysis, paralleling these concepts in spatial statistics (Part I) and spatial econometrics (Part II) work. The principal concept is morphism, which may be defined as a structure preserving functional linkage between mathematical properties or operations in spatial statistics/econometrics and in other disciplines. In other words, conceptions in both domains are structurally the same even though their labeling and notation for their elements differ.

Three categories of morphism are of interest in this book: isomorphisms, homomorphisms, and allomorphisms. An isomorphism is a one-to-one correspondence (i.e., a functional linkage between selected properties such that every property in one domain is paired with one and only one property in the other domain) between the properties of spatial statistics/econometrics and another discipline such that the result of an operation yields an analogous result in both domains. A homomorphism is a transformation from another discipline into spatial statistics/econometrics that allows the modeling, simulation, or representation of the structure of the origin discipline within spatial statistics/econometrics. In other words, a similarity of form occurs between pairs of unrelated objects from some other discipline and from spatial statistics/econometrics. Together isomorphisms and homomorphisms reflect on model specification similarity between other disciplines and spatial statistics/econometrics. Finally, an allomorphism occurs when a model specification appearing in both another discipline and spatial statistics/econometrics is similar in composition but different in form. In other words, spatial statistics/econometrics adopts specifications from other disciplines.

The chapters of this book use these morphisms as organizing concepts in the context of different empirical and conceptual applications. Employment data furnish an empirical basis for the presented spatial statistics (Part I).

D. A. Griffith, J. H. P. Paelinck, *Morphisms for Quantitative Spatial Analysis*, Advanced Studies in Theoretical and Applied Econometrics 51,
https://doi.org/10.1007/978-3-319-72553-6_1

1.2 Polish Employment Data: 2006–2013

The empirical data used for illustrative purposes comprises the geographic distribution of annual employment by five general categories across Poland for the powiat (NUTS4 $\approx$ county; $n = 380$; see Fig. 1.1a) spatial resolution during 2006–2013. Covariates in this description are working age population density (Fig. 1.1c) and percentage of working age population (Fig. 1.1d).

Employment data were obtained from the Polish Statistical Office, which classified them as follows:

A: agriculture, forestry, hunting & fishing
I: industry and construction

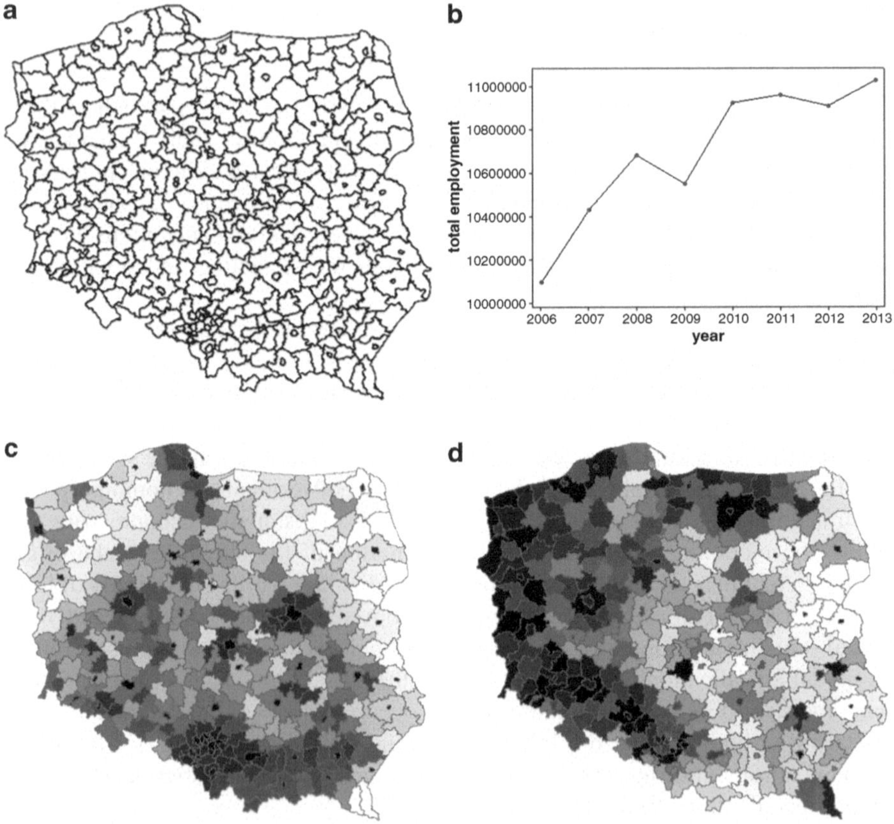

Fig. 1.1 Top left (**a**): The powiaty of Poland for 2006–2013. Top right (**b**): time-series plot of total national Polish employment for 2006–2013. Bottom left (**c**): working age population density principal component. Bottom right (**d**): percentage of working age population principal component. The geographic information system (GIS) files were obtained from *GIS Support Sp. Z oo* and can be downloaded from the webpage http://www.codgik.gov.pl/index.php/darmowe-dane/prg.html (last accessed on 21 October 2017)

T: trade; repair of motor vehicles; transportation and storage; accommodation and catering; information & communication
S: financial & insurance activities; real estate services
O: other services

1.3 Polish Data Quality

The space–time dataset comprises 8 time periods and 380 polygon areal units. Total employment does not equal the sum of the employment categories for the years and numbers of areal units appearing in Table 1.1.

In addition, Wałbrzych (a city of 100,000+ inhabitants) has no employment in any category for the years 2006–2012 (otherwise, the smallest total employment is 4290).

An inspection of powiaty time series containing 0 entries implies that the 4 years with excess total employment actually should have the residual employment allocated to those few powiaty containing 0 entries. This allocation was done according to the percentages in the employment classes immediately prior to and/or immediately following the 0 entry years. The resulting allocations are consistent with the individual powiat time series and hence seem sensible. Table 1.2 reports their tabulation (gray denotes data known).

Meanwhile, in 2013, Wałbrzych became a separate powiat from its hinterland powiat. The allocation for this year is 67.1% for the city (Powiat #334) and 32.9% for its hinterland (Powiat #379). Applying these percentages produces the allocation of population summarized in Table 1.3.

The accompanying 2013 employment category percentages render the employment count imputations appearing in Table 1.4.

All of these imputations were included for the data analyses summarized in this part of the book.

Of note is that employment appears to be model—rather than survey-based. Evidence supporting this contention is that the working age population reduces to a single principal component across the 8 years that accounts for 100% of its variance, implying that these quantities are perfect linear combinations of some set of benchmark measures.

Finally, these data still contain two minor discrepancies. For 2007, the total population for Powiaty #344 and #379 differ from their sums of preworking, working, and post-working age populations by, respectively, 1 and −28.

Table 1.1 Frequency, by year, of total employment not equaling the sum of the employment category counts

Year	Number of powiaty
2006	4
2007	2
2009	14
2010	8

Table 1.2 Imputations of missing values by powiat and year

Powiat ID	year	A	I	T	S	total
112	2006		839		27	866
118	2006		1385		129	1514
328	2006	149			537	686
331	2006	98			822	920
328	2007	155			505	660
331	2007	105			728	833
11	2009	384			673	1057
13	2009	269			790	1059
191	2009	10092			94	10186
229	2009	71			2094	2165
241	2009	15529			415	15944
275	2009	91		1918		2009
276	2009	50		2233		2283
278	2009	83		8145		8228
279	2009	620			473	1093
304	2009	1116			8256	9372
322	2009	110			1001	1111
330	2009	109			864	973
369	2009	676			6969	7645
371	2009	567			918	1485
13	2010	1396			745	2141
14	2010	1785			545	2330
259	2010	4758			106	4864
275	2010	239			930	1169
278	2010	219		8692		8911
279	2010	728		2787		3515
286	2010	441			1170	1611
330	2010	337			829	1166

Table 1.3 Disaggregation population imputation for the Wałbrzych city and hinterland powiaty

Year	Powiat #344	Powiat #379	Total
2006	123,246	60,443	183,689
2007	122,183	59,921	182,104
2008	121,291	59,485	180,776
2009	120,453	59,073	179,526
2010	120,911	59,298	180,209
2011	119,950	58,826	178,776
2012	119,163	58,441	177,604

1.4 Concluding Comments

Organizing concepts for this book include morphisms, which each chapter highlights. The spatial statistical treatment uses Polish employment data that have undergone careful cleaning and editing described in this chapter. The substantive

Table 1.4 2013 employment category count imputations for the Wałbrzych city and hinterland powiaty

Powiaty ID	Year	A	I	T	F	S	Total
344	2006	272	13,725	5297	1237	10,579	31,110
344	2007	268	14,857	5293	1253	11,332	33,003
344	2008	272	14,905	5781	1271	11,117	33,346
344	2009	230	12,303	5679	1143	11,075	30,430
344	2010	317	12,075	5889	1014	11,024	30,319
344	2011	316	11,945	5665	960	11,087	29,973
344	2012	309	11,716	5372	953	11,016	29,366
379	2006	1104	2138	1352	272	2104	6970
379	2007	1088	2314	1351	276	2254	7283
379	2008	1104	2322	1476	279	2211	7392
379	2009	934	1916	1450	251	2203	6754
379	2010	1288	1881	1503	223	2193	7088
379	2011	1285	1861	1446	211	2205	7008
379	2012	1256	1825	1371	210	2191	6853

chapters in both Parts I and II address parallel themes in the same sequential order. Chapter 2 discusses spatial autocorrelation and the location–allocation problem. The spatial statistical content of this chapter parallels the spatial econometrics context of Chap. 13 in terms of spatial optimization, emphasizing morphisms pertaining to combinatorial analysis and metric typology. Chapter 3 addresses space–time autocorrelation. The spatial statistical content of this chapter parallels the spatial econometrics context of Chap. 14 in terms of spatial time series, emphasizing morphisms pertaining to econotimespace, forging a close integration of space–time and empirical economics (Chap. 3)/economic theory (Chap. 14). Chapter 4 treats spatially structured and unstructured random effects terms. The spatial statistical content of this chapter parallels the spatial econometrics context of Chap. 15 in terms of spatial dynamics and control, Chap. 15 emphasizing morphisms in terms of operations research, and Chap. 4 in terms of time-series analysis. Chapter 5 describes relationships between eigenvector spatial filters and the spatial weights matrix. The spatial statistical content of this chapter parallels the spatial econometrics context of Chap. 16 in terms of spatial weights matrix specification, emphasizing morphisms as formal bridgings between spatial econometrics/statistics and other fields of investigation. Chapter 6 analyzes location quotients in terms of their spatial autocorrelation components. The spatial statistical content of this chapter parallels the spatial econometrics context of Chap. 17 in terms of spatial clustering, emphasizing morphisms pertaining to generic clusters as well as nonstandard approaches to clustering. Chapter 7 examines estimation for massively large georeferenced datasets. The spatial statistical content of this chapter parallels the spatial econometrics context of Chap. 18 in terms of estimation problems, emphasizing morphisms arising from an invoking of analytical geometry; Chap. 7 does so in order to be able to estimate parameters of spatial models for massive georeferenced datasets forming irregularly shaped regions, whereas Chap. 18 does so to achieve a formal specification of demand theory. Chapter 8 evaluates similarities and differences between fixed and random effects specifications. The spatial statistical content of this chapter parallels the spatial econometrics context of Chap. 19 in terms of spatial structural indicators, emphasizing morphisms pertaining to spatial discount functions and concentration indices (vis-à-vis spatially structured effects). Chapter 9 reviews the state-of-the-art of spatial autocorrelation in spatial interaction data. The spatial statistical content of this chapter parallels the spatial econometrics context of Chaps. 20 and 21 in terms of spatial flows and geographic complexity, emphasizing morphisms pertaining to spatial constraints and geographic clustering associated with movement in space. Finally, Chap. 10 summarizes general conclusions and implications about spatial statistics based upon the content of Part I of this book.

Chapter 2
Spatial Autocorrelation and the *p*-Median Problem

2.1 Introduction

The Weberian location problem involves determining locations in continuous space, either a single location in its simplest form or multiple locations in its more complicated form. Chapter 12 addresses this situation in a TBS context; this chapter addresses it in more of a public goods context. The morphism here continues to relate to combinatorial analysis. Spatial autocorrelation can come into play in two distinct ways: (1) it can be viewed as duplicate information latent in the weights used in minimization and, hence, exploited when missing weight values occur in order to impute them, and (2) it can help inform determination of an optimal solution. This chapter illustrates these notions in terms of $p = 1$ and 2 and a number of simulation experiments. Empirical weights are based upon the geographic distribution of 2013 employment by one of the five general categories across Poland for the geographic resolution of a powiat (NUTS4 $\approx$ county; $n = 380$). This chapter extends the existing literature by implementing an eigenvector spatial filter (ESF) version of the missing data imputation problem and by furnishing timing experiments for the *p*-median solutions.

2.2 Eigenvector Spatial Filtering in a Nutshell

Eigenvector spatial filtering (Griffith 2003a) is a novel spatial statistical methodology that adds a set of synthetic proxy variables, which are eigenvectors extracted from an *n*-by-*n*, usually binary 0-1, spatial weights matrix (see Chap. 5) **C** that links geographic objects together in space, as control variables to filter spatial autocorrelation out of residuals and transfer it to the mean response in a model specification. These control variables identify and isolate the stochastic spatial dependencies

D. A. Griffith, J. H. P. Paelinck, *Morphisms for Quantitative Spatial Analysis*, Advanced Studies in Theoretical and Applied Econometrics 51,
https://doi.org/10.1007/978-3-319-72553-6_2

among georeferenced observations, thus allowing model parameter estimation to proceed with observations mimicking being independent.

The crucial mathematical quantities from matrix **C** are eigenfunctions, which are n pairs of quantities computed via the matrix determinant of a modified version of matrix **C**, **MCM**, $\mathbf{M} = (\mathbf{I} - \mathbf{11}^{\mathrm{T}}/n)$, where **I** denotes the n-by-n identify matrix, and **I** denotes the n-by-1 vector of ones: a scalar (eigenvalue) and a vector (its corresponding eigenvector). Eigenvalues are the n scalar solutions to the nth order polynomial matrix equation det $(\mathbf{MCM} - \lambda\mathbf{I}) = 0$; the corresponding eigenvectors **E** are the nontrivial vector solutions to the equation $(\mathbf{MCM} - \lambda\mathbf{I})\mathbf{E} = \mathbf{0}$. These eigenfunctions are the basis of eigenvector spatial filtering and are the synthetic variates that account for nonzero spatial autocorrelation in spatial regression residuals.

The Moran Coefficient (MC) index of spatial autocorrelation may be written, using matrix notation, for some random variable Y with n georeferenced observations, as

$$\frac{n}{\mathbf{1}^{\mathrm{T}}\mathbf{C}\mathbf{1}} \frac{\mathbf{Y}^{\mathrm{T}}\left(\mathbf{I} - \mathbf{11}^{\mathrm{T}}/n\right)\mathbf{C}\left(\mathbf{I} - \mathbf{11}^{\mathrm{T}}/n\right)\mathbf{Y}}{\mathbf{Y}^{\mathrm{T}}\left(\mathbf{I} - \mathbf{11}^{\mathrm{T}}/n\right)\mathbf{Y}} \tag{2.1}$$

ESFs are constructed as linear combinations of the eigenvectors from matrix

$$\left(\mathbf{I} - \mathbf{11}^{\mathrm{T}}/n\right)\mathbf{C}\left(\mathbf{I} - \mathbf{11}^{\mathrm{T}}/n\right).$$

Appealing properties of these eigenvectors include: (1) they are mutually orthogonal and uncorrelated; (2) one vector is proportional to the vector **1**, the intercept covariate in a regression model; and (3) eigenvalues index and eigenvectors can be used to visualize various natures and degrees of spatial autocorrelation.

Including eigenvectors as covariates, selecting relevant ones with a stepwise procedure enables spatial autocorrelation to be accounted for in a conventional statistical estimation context, in either a linear or a generalized linear model specification.

2.3 Imputing Missing Spatial Data

Griffith (2010) outlines solutions to the aspatial missing data problem. This section presents extensions of those solutions to the spatial missing data problem.

Suppose $\mathbf{Y}_{\mathrm{o}}$ denotes the n_{o}-by-1 ($n_{\mathrm{o}} = n - n_{\mathrm{m}}$) vector of observed response values, and $\mathbf{Y}_{\mathrm{m}}$ denotes the n_{m}-by-1 vector of missing response values. Let $\mathbf{X}_{\mathrm{o}}$ denote the vector of predictor values for the set of observed response values, and $\mathbf{X}_{\mathrm{m}}$ denote the vector of predictor values for the set of missing response values. Let $\mathbf{0}_{\mathrm{o}}$ denote an n_{o}-by-n_{m} matrix of zeroes, and $\mathbf{0}_{\mathrm{m}}$ denote an n_{m}-by-1 vector of zeroes. Let **1** denote an n-by-1 vector of ones that can be partitioned into $\mathbf{1}_{\mathrm{o}}$, denoting the vector of ones for the set of observed response values, and $\mathbf{1}_{\mathrm{m}}$, denoting the vector of ones for the set of missing response values. Let $\mathbf{I}_m$ denote an n_m-by-n_m identity matrix. Finally, let $\mathbf{W}$ denote the row-standardized spatial weights matrix, with $\mathbf{W}_{oo}$ containing the spatial weights for the pairs of known value locations, $\mathbf{W}_{om}$ and $\mathbf{W}_{mo}$ containing the spatial weights for the pairs of know with unknown value locations, and $\mathbf{W}_{mm}$ containing the spatial weights for the pairs of unknown value locations. The spatial statistical simultaneous autoregressive (SAR) model may be written, using partitioned matrix notation, for imputation purposes as

$$\begin{pmatrix} \mathbf{Y}_o \\ \mathbf{0}_m \end{pmatrix} = \beta_0 \begin{pmatrix} \mathbf{1}_o \\ \mathbf{1}_m \end{pmatrix} + \begin{pmatrix} \mathbf{X}_o \\ \mathbf{X}_m \end{pmatrix} \boldsymbol{\beta} + \rho \begin{pmatrix} \mathbf{W}_{oo} & \mathbf{W}_{om} \\ \mathbf{W}_{mo} & \mathbf{W}_{mm} \end{pmatrix} \begin{pmatrix} \mathbf{Y}_o \\ \mathbf{Y}_m \end{pmatrix} + \begin{pmatrix} \mathbf{0}_o \\ -\mathbf{I}_m \end{pmatrix} (\mathbf{Y}_m) + \boldsymbol{\varepsilon}, \tag{2.2}$$

where β_0 is the intercept term, $\boldsymbol{\beta}$ is the p-by-1 vector of regression coefficients for p covariates, ρ is the spatial autocorrelation parameter, and $\boldsymbol{\varepsilon}$ is an n-by-1 vector of iid normal random errors. In this specification, $\mathbf{Y}_m$ is treated as a vector of parameters during estimation (after all, it is a vector of conditional expectations), which requires nonlinear regression techniques. As such, $\boldsymbol{\varepsilon}_m \equiv 0_m$. These imputations are those discussed by Griffith et al. (1989) and are equivalent to those obtained with kriging (Griffith and Layne 1997).

Not only does the estimation of Eq. (2.2) involve a Jacobian term, but this term is more complex than the standard spatial statistical Jacobian term. It is given by

$$-\frac{2}{n - n_m} \left[\sum_{i=1}^{n} \mathrm{LN}(1 - \rho\lambda_i) - \sum_{k=1}^{n_m} \mathrm{LN}(1 - \rho\omega_k) \right],$$

where the first set of n eigenvalues (λ_i) is for matrix $\mathbf{W}$, and the second set of m eigenvalues (ω_k) is for matrix $\mathbf{W}_{mm}$.

Eigenvector spatial filtering circumvents the preceding complications, which are more severe for non-normal random variables (e.g., Poisson and binomial). It replaces the spatial weights matrix components in Eq. (2.2) with an eigenvector spatial filter (ESF). This ESF—which is a linear combination of selected eigenvectors—specification permits the calculation of standard imputations for a linear model. The estimation equation here for a linear model specification is

$$\begin{pmatrix} \mathbf{Y}_o \\ \mathbf{0}_m \end{pmatrix} = \beta_0 \begin{pmatrix} \mathbf{1}_o \\ \mathbf{1}_m \end{pmatrix} + \begin{pmatrix} \mathbf{X}_o \\ \mathbf{X}_m \end{pmatrix} \boldsymbol{\beta}_X + \begin{pmatrix} \mathbf{0}_o \\ -\mathbf{I}_m \end{pmatrix} (\mathbf{Y}_m) + \sum_{k=1}^{K} \begin{pmatrix} \mathbf{E}_{o,k} \\ \mathbf{E}_{m,k} \end{pmatrix} \beta_{E_K} + \boldsymbol{\varepsilon}, \tag{2.3}$$

where $\boldsymbol{\beta}_X$ denotes the p-by-1 vector of regression coefficients for p covariates, K is the number of eigenvectors selected from the candidate set from which an ESF is constructed, $\mathbf{E}_{o,k}$ and $\mathbf{E}_{m,k}$, respectively, denote the parts of eigenvector k associated with the observed and missing values, and β_{E_K}denotes the regression coefficient for eigenvector $\mathbf{E}_k$ used to construct the ESF in question. Imputations in this chapter are based upon the GLM counterpart to Eq. (2.3), which is one of the contributions of this chapter to the literature [earlier work is based upon Eq. (2.2)].

2.4 The Location–Allocation Problem

The p-median problems analyzed here are for $p = 1$ and 2. They are the Weberian location problem of spatial economics (Weber 1909), and its solution is the locations of P central facilities serving n demand points such that the cost of flows between each demand point and its closest central facility is a minimum. Scott (1970) furnishes a review of the early literature about this problem; Farahani and Hekmatfar (2009), and Eiselt and Marianov (2011) furnish updated overviews of this literature. This spatial optimization problem may be stated formally as follows:

$$\text{MIN} : Z = \sum_{j=1}^{P} \sum_{i=1}^{n} \lambda_{ij} w_i \sqrt{\left(u_i - U_j\right)^2 + \left(v_i - V_j\right)^2} \tag{2.4}$$

$$\text{s.t.} : \sum_{j=1}^{P} \lambda_{ij} = 1 \forall i$$

$$1 \le \sum_{i=1}^{P} \lambda_{ij} < n \;\; \text{for}\, j = 1, 2$$

where

(u_i, v_i) are the Cartesian coordinates of demand point i;
w_i is the weight of demand point i (here it is a LQ);
(U_j, V_j) are the Cartesian coordinates of central facility j; and
$\lambda_{ij} = \begin{cases} 1, \text{if demand point } i \text{ is assigned to central facility } j \\ 0, \text{otherwise} \end{cases}$

This specification ensures that each demand point is allocated to one and only one central facility, and that each central facility has a least one demand point allocated to it. Figure 2.1b portrays the geographic distribution of w_is ($i = 1, 2, \ldots, 380$) for analyses summarized in this chapter.

For $p = 1$, the constraints disappear, and the objective function reduces to

$$\text{MIN} : Z = \sum_{i=1}^{n} w_i \sqrt{\left(u_i - U_j\right)^2 + \left(v_i - V_j\right)^2},$$

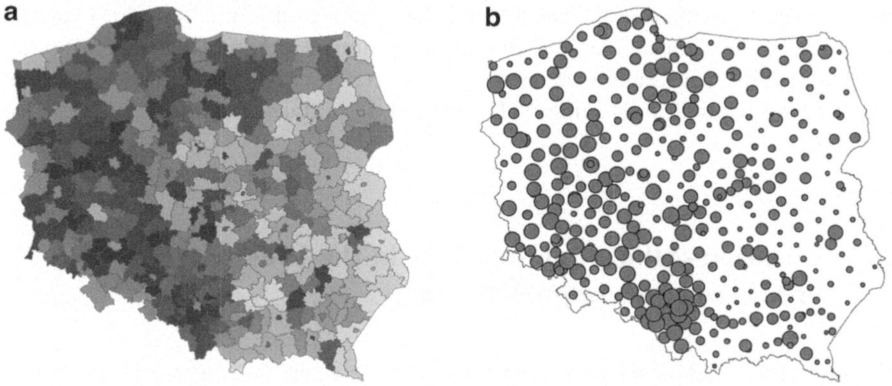

Fig. 2.1 Geographic distribution of 2013 industrial employment location quotients (LQs) across Poland. Left (**a**): by powiat; LQs are directly proportional to the darkness of the grayscale. Right (**b**): converted to weights and attached to powiat centroids

which has a simple algorithmic (i.e., automated) solution established by Kuhn and Kuenne (1962) based upon the derivatives of the objective function expression. For $p = 2$, the objective function becomes

$$\text{MIN} : Z = \sum_{j=1}^{2} \sum_{i=1}^{n} \lambda_{ij} w_i \sqrt{\left(u_i - U_j\right)^2 + \left(v_i - V_j\right)^2}.$$

The goal here is to determine two central facility locations and then allocate each of the demand points to its closest central facility such that the sum of the weighted distances from each demand point to its closest central facility is a minimum. In contrast with the $p = 1$ problem, now not only do central facility locations need to be determined, but the allocation of points also needs to be determined. The number of possible feasible allocations of n demand points to two central facilities is $n(n-1)/2$. Ostresh (1975) furnishes an efficient algorithm for solving this problem. His algorithm is based upon a geometric solution that uses a line to partition the geographic landscape into two regions, and then rotates this line, yielding an exact solution.

The combinatorial nature of location–allocation problems results in a massive number of possible feasible solutions, often too many to evaluate in a reasonable amount of time, let alone almost instantly. Accordingly, an exact solution may not be obtainable in an automated way. Heuristic algorithms have been developed as a response to this challenge. A heuristic algorithm is based upon selected principles that allow the algorithm to quickly find a good, but not necessarily optimal, solution. ALTERN (Cooper 1964) is a popular heuristic for location–allocation problems. It exploits the following two principles: (1) if the allocation of n demand points to central facilities is known, then the $p = 1$ problem can be solved for each grouping of demand points, and (2) if the locations of central facilities are known, then n demand points can be allocated to their closest central facility. This heuristic alternates between these two principles until the sum of weighted distance no longer decreases. This sum cannot decrease below that for the optimal solution. The heuristic is sensitive to its initial solution, and an initial solution often is selected at random.

2.5 Location–Allocation Solutions in the Presence of Missing and Imputed Data

Griffith (1997, 2003b) addresses the impacts of missing data and use of its spatial statistical imputations on solutions to p-median problems, where $p = 1, 2, 3$. These results are replicated in this section, but for $p = 1$ and using an ESF model specification and Polish powiat employment data, specifically, the geographic distribution of 2013 industrial location quotients (LQs; Fig. 2.1a). Figure 2.1b portrays the geographic distribution of weights (proportional to the LQs, which range from 0.1 to 2.7), which are linked to powiat centroids to constitute demand points, and they are used in the subsequent location–allocation simulation experiments.

Utilizing these data, cross-validation and bootstrap samples were generated to reproduce the earlier imputation experiments. Because Poland has 380 powiaty, the number of samples was set to 380 to maintain comparability between the cross-validation and bootstrap findings. Table 2.1 summarizes results from this exercise. Using a 10% level of significance, the constructed ESF, which has a $MC/MC_{max} = 0.99/1.175 = 0.84$, contains 22 eigenvectors[1] (selected from a candidate set of 85 for which each has a $MC/MC_{max} \geq 0.25$) that account for nearly half of the variation in these LQs (indicating the presence of moderate positive spatial autocorrelation; the accompanying $MC = 0.45$). The number of imputations ranges from 1 to 76 (20% of the number of powiaty). ESFs account for roughly 30% of the geographic variability in the imputed industrial LQs across all of these random suppressions (Fig. 2.2 displays specimen scatterplots of them). Five of the eigenvectors are selected in all cases; 17 of the eigenvectors are selected for either one or two missing values. The prominence of eigenvectors fails to relate to their level of significance in the original model without missing values: Pr $(E_{15}) < 0.0001$ but becomes compromised immediately, with $n_m = 1$, whereas Pr $(E_{58}) = 0.0702$ does not become compromised until $n_m > 19$.

As reported in previous analyses, imputations result in a concentration of the $p = 1$ solution. Figure 2.3a reveals how the 380 different solutions cluster, for both $n_m = 1$ and $n_m = 76$. Figure 2.3b, c reveal greater scatter with an increasing number

[1]A beta-binomial probability model was employed to account for extra-binomial variation in the geographic distribution of LQs. The basic effect of using this parametric mixture distribution is an increase in standard errors, and hence a reduction in the number of selected eigenvectors.

Table 2.1 Cross-validation/bootstrap ESF and imputation results for 380 replicates

Eigenvector	$n_m = 0$ H_0 prob	$n_m = 0$	$n_m = 1$	$n_m = 2$	$n_m = 3$	$n_m = 4$	$n_m = 5$	$n_m = 10$	$n_m = 19$	$n_m = 38$	$n_m = 76$
E_1	<0.0001	1	380	380	380	380	380	380	380	380	380
E_2	<0.0001	1	380	380	380	380	380	380	380	380	380
E_3	<0.0001	1	380	380	380	380	379	371	348	313	247
E_6	0.0377	1	380	380	380	380	380	380	380	377	362
E_8	0.0042	1	380	380	380	380	380	380	380	380	380
E_9	0.0002	1	380	380	380	380	380	380	380	380	380
E_{15}	<0.0001	1	379	377	376	375	370	351	310	255	207
E_{18}	0.0455	1	380	380	380	380	380	380	380	369	344
E_{19}	0.0143	1	380	380	380	380	380	380	380	380	380
E_{22}	<0.0001	1	380	380	379	379	376	372	352	315	227
E_{25}	0.0332	1	380	380	380	380	379	374	364	324	266
E_{26}	0.0313	1	331	296	292	274	254	234	214	197	167
E_{27}	0.0924	1	339	292	265	241	243	187	169	153	116
E_{44}	0.0882	1	380	380	380	380	380	380	378	373	316
E_{46}	0.0122	1	366	350	320	314	287	248	206	182	152
E_{58}	0.0702	1	380	380	380	380	380	380	380	368	322
E_{59}	0.0137	1	380	380	380	380	380	380	379	357	305
E_{65}	0.0147	1	380	380	380	380	380	380	378	375	339
E_{68}	0.0118	1	380	380	380	380	380	377	360	333	263
E_{70}	0.0368	1	380	380	380	380	380	380	380	380	361
E_{75}	0.0048	1	380	380	380	380	380	380	380	380	374
E_{82}	0.0013	1	366	343	329	300	299	246	224	180	152
Others		0	72	147	238	293	349	511	793	1057	1391
Average #		0	0.19	0.39	0.63	0.77	0.92	1.34	2.09	2.78	3.66

(continued)

Table 2.1 (continued)

Eigenvector	$n_m = 0$ H_0 prob	$n_m = 0$	$n_m = 1$	$n_m = 2$	$n_m = 3$	$n_m = 4$	$n_m = 5$	$n_m = 10$	$n_m = 19$	$n_m = 38$	$n_m = 76$
Intercept		0.16	2.12	1.50	2.73	2.06	2.24	2.11	2.34	2.38	2.65
Slope		0.98	0.76	0.82	0.67	0.77	0.74	0.76	0.73	0.72	0.70
Pseudo-R^2		0.48	0.31	0.35	0.28	0.32	0.31	0.31	0.30	0.30	0.28

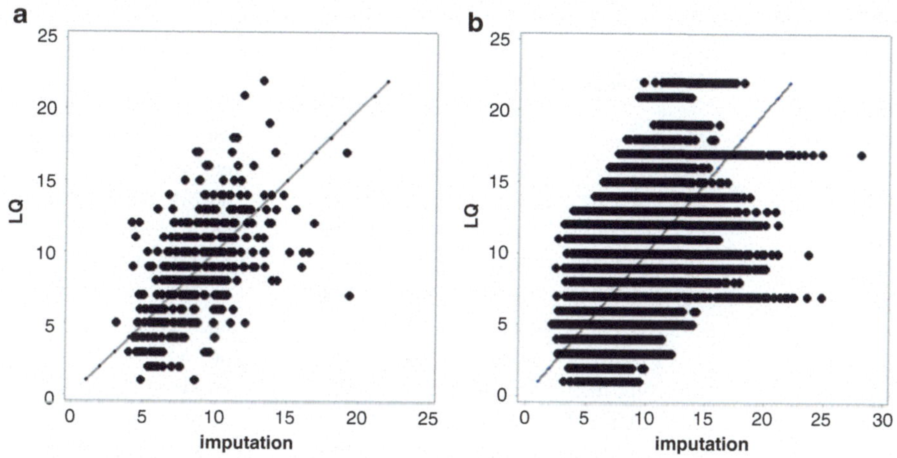

Fig. 2.2 Scatterplots of the observed LQs versus their imputations. Left (**a**): $n_m = 1$. Right (**b**): $n_m = 76$

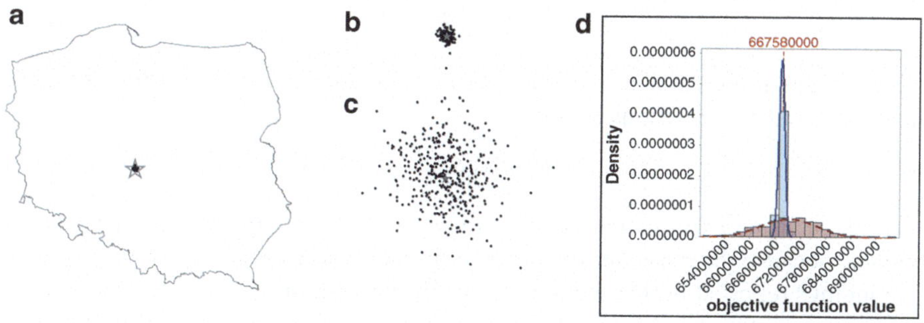

Fig. 2.3 The geographic distribution of $p = 1$. Left (**a**): the exact solution, denoted by a gray star, superimposed on the 380 $n_m = 1$ and $n_m = 76$ imputation solutions. Top middle (**b**): a zoom in for the $n_m = 1$ solutions. Bottom middle (**c**): a zoom in for the $n_m = 76$ solutions. Right (**d**): histograms of objective function values for $n_m = 1$ and $n_m = 76$, with the actual objective function value superimposed

of imputations. Figure 2.3d reveals an increasing spread of the objective function values with an increasing number of imputations; these frequency distributions closely conform to a bell-shaped curve and essentially are symmetrically distributed around the actual objective function value.

Table 2.2 summarizes comparison results for the varying number of missing values. The ESF imputations result in considerable stability. The average (U, V) coordinates across numbers of missing values are remarkably stable, as is the mean objective function value. The standard distance increases, as expect, as does the standard deviation of the objective function value, with an increasing number of

Table 2.2 Summary measures of imputation impacts on the $p = 1$ location–allocation solution

Number of imputations	Centrographic measures			Objective function ($\times 10^4$)	
	Mean u	Mean v	Standard distance	Mean	Standard deviation
$n_m = 0$	483,218.41194	426,658.94972	0	66,758	0
$n_m = 1$	483,219.05677	426,663.58251	280.65	66,761	68.71
$n_m = 2$	483,212.90951	426,674.20923	383.98	66,767	96.95
$n_m = 3$	483,198.89484	426,676.34528	486.36	66,771	121.64
$n_m = 4$	483,222.29443	426,684.93152	555.06	66,762	142.94
$n_m = 5$	483,255.99839	426,672.67385	628.52	66,772	154.32
$n_m = 10$	483,232.40197	426,649.81399	933.94	66,769	215.70
$n_m = 19$	483,250.01022	426,704.22133	1230.55	66,816	304.60
$n_m = 38$	483,280.04454	426,740.53942	1840.06	66,868	487.53
$n_m = 76$	483,240.53114	426,912.26792	2719.89	66,938	686.87

imputations. The increase trend for the standard distance is a power function of the following form:

$$262\sqrt{n_m},$$

which means it increases at a decreasing rate; the increase trend for the objective function standard deviations is similar:

$$64\sqrt{n_m}.$$

Both of these expressions yield 0 for $n_m = 0$. These trends also have upper bounds, because the maximum number of missing values is $n-1$.

Findings reported in this section are consistent with previous ones. One of their implications is that an ESF imputation can be just as effective as a SAR imputation. Another is that imputations based upon spatial autocorrelation alone can be relatively simple to calculate and effective in preserving a good location–allocation solution, without having to resort to other predictor covariates (e.g., working age population density). Of course, relevant covariates improve an imputation (e.g., co-kriging), but they may not be essential in spatial analysis.

2.6 Relationships Between Spatial Autocorrelation and Solutions to Location–Allocation Problems

Griffith and Chun (2015) explore relationships between spatial autocorrelation and location–allocation problem exact solutions, noting a tendency for optimal locations to coincide with local indices of spatial autocorrelation (LISA) hot spots. This section summarizes timing experiments as well as heuristic solution trends for location–allocation problems with $p = 1$ and 2 using an ESF to simulate the geographic distribution of weights.

Graham (1995) notes that a relationship exists between the intercept term and the spatial autocorrelation parameter in an auto-binomial model specification. The ESF specification has this same property. In the Polish industrial LQs case, this relationship may be described as follows (a stable model specification in spatial statistical parlance):

$$\text{intercept} = \alpha - 1.60\left(1 - e^{-0.03\beta^{1.82}}\right),$$

where α is the intercept parameter of a nonspatial binomial random variable, and β is the coefficient attached to the ESF. This intercept term introduces the random component, whereas the ESF introduces the map pattern component, of a simulation experiment. The mean LQ for the simulation experiments is 8.43 (the intercept term is −2.38540), with a standard deviation of 2.78. Each simulation experiment involves 10,000 random draws from a binomial distribution with an intercept of $-2.38540 - 1.60\left(1 - e^{-0.03\beta^{1.82}}\right)$ and β values of 3.5, 1.0, and 0.4; the 1.0 case is equivalent to the empirical LQ case. Table 2.3 summarizes features of these data. The pseudo-R^2 values index the percentage of variance accounted for by the ESF, indicating sets of surfaces containing weak, containing moderate, and containing strong spatial autocorrelation. The average map values are very similar to that for the original Poland map. And, the variance inflates with increasing positive spatial autocorrelation.

Figure 2.4 presents maps summarizing the simulated data for high positive spatial autocorrelation. A visual comparison of Figs. 2.1a and 2.4a, the geographic distribution of the mean simulated industrial LQs, confirms that, on average, the simulated data retain the underlying map pattern for the LQs. Figure 2.4b portrays the variability for the simulated LQs. Because LQs essentially are binomial random variables, their means and variances are related, which Fig. 2.4b reflects (it has the same map pattern as Fig. 2.4a). These outcomes illustrate another advantage of ESF over spatial auto-models: preservation of map pattern. Figure 2.4c is the average of 10,000 replications generated with an auto-binomial model containing markedly high positive spatial autocorrelation (the equivalent of $\rho = 0.9$ for an SAR model). The ESF map is based on average values ranging from 0 to 61, whereas the auto-binomial map is based on average values ranging from 45 to 48; essentially any map pattern has been average out of this latter composite map. Furthermore, Fig. 2.4d reveals that a standard deviation in the auto-model map is not strongly proportional

Table 2.3 Summary statistics for the simulated sets of weights, 10,000 replications

β	pseudo − R^2	s_{R^2}	intercept	$\overline{\overline{y}}$	$s_{\bar{y}}$	$\bar{y}_{min}$	$\bar{y}_{max}$	$\overline{s}_y$
3.5	0.90908	0.00851	−2.78263	8.93993	0.14078	8.41316	9.55526	9.00026
1.0	0.44334	0.03447	−2.43317	8.40154	0.14111	7.79474	8.96053	3.70101
0.4	0.11718	0.02980	−2.39507	8.41169	0.14093	7.93421	8.96316	2.94895

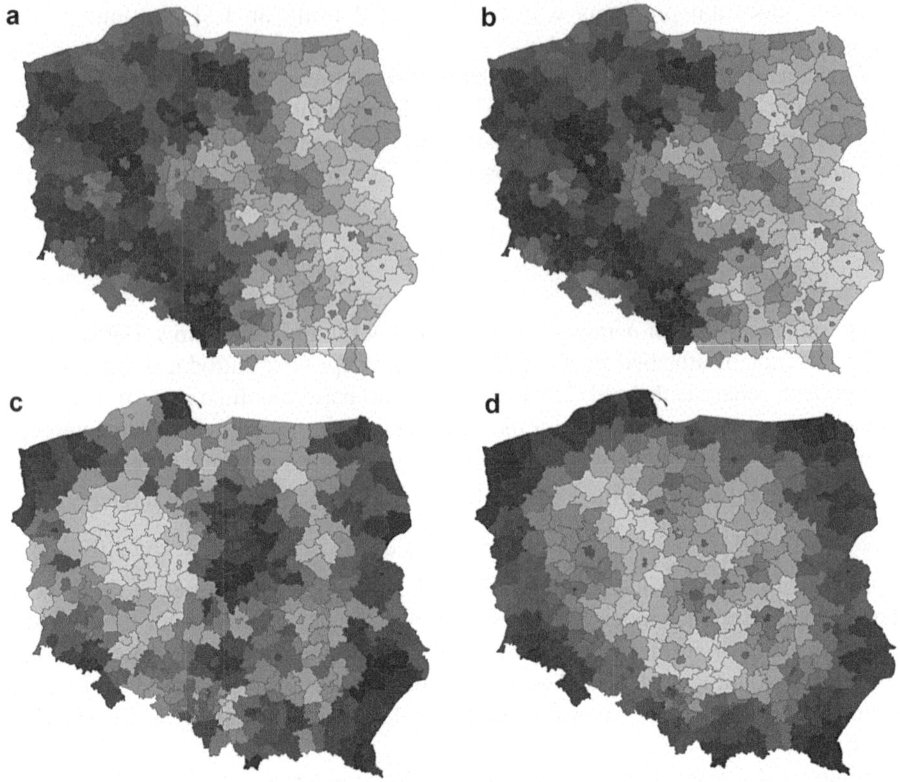

Fig. 2.4 Summary maps for simulated data based on 10,000 replications; the grayscale is directly proportion to the magnitude of LQ means and standard deviations. Top left (**a**): geographic distribution of the areal unit ESF-based simulation means. Top right (**b**): geographic distribution of the areal unit ESF-based standard deviations. Bottom left (**c**): geographic distribution of the auto-binomial-based areal unit simulation means. Bottom right (**d**): geographic distribution of the auto-binomial-based areal unit standard deviations

to its corresponding mean. Rather, Fig. 2.4d reflects the differences in variances in the center and along the edges of a geographic landscape (this in part arises from the **W** matrix). The ESF map standard deviations (Fig. 2.4b) range from 0.4 to 4.9, whereas the auto-binomial map standard deviations (Fig. 2.4d) range from 28 to 36. Consequently, ESF-based simulations are preferable to the auto-model simulations.

Figure 2.5 presents local spatial autocorrelation indices for the high positive spatial autocorrelation ESF map ($\beta = 3.5$): Fig. 2.5a portrays the geographic distribution of local MC_is, and Fig. 2.5b portrays the geographic distribution of local G_is. The $p = 1$ solution begins with the spatial mean, resulting in an average of six iterations to determine the optimal solution, with a range of 5–9 iterations. Both the maximum G_i and the maximum MC_i values identify the same initial location, which results in an average of eight iterations, with a range of 6–13; only 18 of the

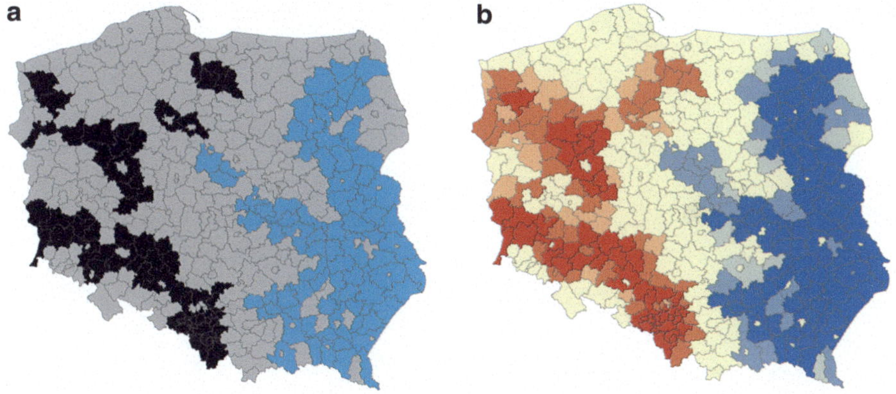

Fig. 2.5 Spatial autocorrelation-based clusters. Left (**a**): local indices of spatial autocorrelation (LISA) extremes (black denotes HH clusters; blue denotes LL clusters). Right (**b**): G_i extremes (red denotes high value hot spots; blue denotes low value hot spots)

Table 2.4 Number of iterations for the Kuhn–Kuenne algorithm, 10,000 replications

Initial location	Weak positive SA			Moderate positive SA			Strong positive SA		
	Mean	Range	# < SM	Mean	Range	# < SM	Mean	Range	# < SM
SM	3.8	3–4		4.1	3–6		5.7	5–9	
MC_i & G_i	4.9	4–7	0	5.1	4–6	5	7.9	6–13	18

Note: SA denotes spatial autocorrelation, SM denotes spatial mean, and MC_i and G_i are the maximums of those indices

10,000 cases have fewer iterations than for the spatial mean initialization. The weaker levels of spatial autocorrelation exhibit similar patterns (Table 2.4). Furthermore, as the degree of spatial autocorrelation increases, more iterations tend to be needed for convergence, and the chances increase that a very few cases benefit from the change in the initial starting location. This outcome most likely relates to the standard relationship existing between means and medians (e.g., they coincide for a bell-shaped curve); Fig. 2.6a reveals that the central location shifts toward the center of Poland (where the mean and the median would coincide) as randomness increasingly dominates the weights. Consequently, exploiting spatial autocorrelation does not appear to improve timing results for the $p = 1$ problem.

The solution for the $p = 2$ case[2] has two central facility locations (Fig. 2.6b) that coincide with the two largest local spatial autocorrelation clusters, and an allocation of the powiaty into two roughly equal size groups. Because determination of an exact solution for each location–allocation simulated set of weights here involves 72,010 solutions, each involving, on average, as many as 6–7 (weak or moderate

[2]To avoid computational difficulties, 1 was added to all weights in order to eliminate 0 weights. These 0 weights are not computationally problematic for the $p = 1$ algorithm solution.

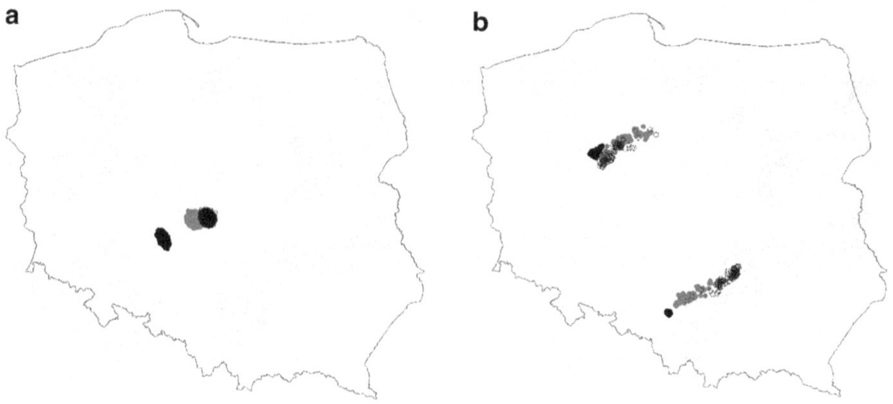

Fig. 2.6 Geographic distribution of simulated map solutions; solid black circles denote strong, solid gray circle denotes moderate, and open circles denote weak positive spatial autocorrelation cases. Left (**a**): $p = 1$. Right (**b**): $p = 2$

Table 2.5 Output from ALTERN for LISA-based and random initializations, 100 replicates

Comparative features	Weak positive SA	Moderate positive SA	Strong positive SA
Initialization based upon LISAs (MC_is)			
# case with fewer iterations	43	97	100
Mean # iterations	10.3	6.6	5.3
Average distance from first central facility	6990.2	6217.4	1139.9
Average distance from second central facility	6815.6	4593.1	408.0
Objective function ratio	0.99983	0.99987	0.99997
Initialization based upon a random allocation			
# case with fewer iterations	57	3	0
Mean # iterations	10.1	15.1	8.8
Average distance from first central facility	6990.2	6217.4	1139.9
Average distance from second central facility	9350.1	11,726.5	452.8
Objective function ratio	0.99974	0.99949	0.99996

Note: SA denotes spatial autocorrelation

spatial autocorrelation) and 8–9 (strong spatial autocorrelation) Kuhn–Kuenne algorithm iterations, only 100 of the simulated cases are analyzed here for each degree of spatial autocorrelation.

Table 2.5 presents summary statistics for the simulation experiments. In virtually all cases, initializing the ALTERN algorithm with LISA-based locations renders better results than initializing it with a random allocation. Results obtained with the LISA-based initializations improve as spatial autocorrelation increases. Both initializations tend to identify a central facility location close to the same optimal location. The LISA start outperforms a random start in identifying the other optimal location. The LISA-based initialization improves timing (i.e., fewer iterations) and renders solutions whose objective function values are closer to the optimal solution objective function value (although both ratios are very close to 1). Given the preceding $p = 1$ results, benefits here must accrue from a better initial allocation (the second principle of the heuristic). Future research needs to address $p > 2$ to evaluate the more general gains that exploiting spatial autocorrelation may be able to furnish for much more difficult location–allocation problems.

2.7 Concluding Comments

Spatial autocorrelation interfaces with location–allocation problems in three ways, one of which is newly reported about in this chapter. The first is when missing weights need to be imputed, a relation known for several decades now, but bolstered in this chapter by couching it in terms of eigenvector spatial filtering. The second is the impact on optimal central facility locations by the way spatial autocorrelation smooths a geographic distribution of weights, a relation known for several years now, but bolstered in this chapter by systematically inspecting increasing degrees of positive spatial autocorrelation embedded in simulated georeferenced data with an empirically based ESF. The third is the relationship between maximum values of local spatial autocorrelation indices (e.g., LISA MC_is and G_is) and the search for optimal central facility locations with heuristics, a topic at the frontiers of spatial autocorrelation research. Of note here is that presently the LISA approach requires a manual determination of locations, especially for $p > 1$, because the desired values are for positive spatial autocorrelation and must be in different clusters for $p > 1$.

References

Cooper, L. (1964). Heuristic methods for location-allocation problems. *SIAM Review, 6*(1), 37–53.

Eiselt, H., & Marianov, V. (Eds.). (2011). *Foundations of location analysis*. New York: Springer.

Farahani, R., & Hekmatfar, M. (Eds.). (2009). *Facility location: Concepts, models, algorithms and case studies*. Berlin: Physica-Verlag.

Graham, J. (1995). Markov chain Monte Carlo inference procedures for discrete spatial lattice models, unpublished doctoral dissertation. Raleigh, NC: Department of Statistics, North Carolina State University.

Griffith, D. (1997). Using estimated missing spatial data in obtaining single facility location-allocation solutions. *l'Espace Géographique, 26*, 173–182.

Griffith, D. (2003a). *Spatial autocorrelation and spatial filtering: Gaining understanding through theory and scientific visualization*. Berlin: Springer.

Griffith, D. (2003b). Using estimated missing spatial data with the 2-median model. *Annals of Operations Research, 122*, 233–247.

Griffith, D. (2010). Some simplifications for the Expectation-Maximization (EM) algorithm: The linear regression model case. *InterStat*, March article 2, 23 pp. Accessed October 15, 2017, from http://interstat.statjournals.net/YEAR/2010/articles/1003002.pdf.

Griffith, D., & Chun, Y. (2015). Spatial autocorrelation in spatial interactions models: Geographic scale and resolution implications for network resilience and vulnerability. *Networks and Spatial Economics, 15*, 337–365.

Griffith, D., & Layne, L. (1997). Uncovering relationships between geo-statistical and spatial autoregressive models. In *1996 Proceedings on the Section on Statistics and the Environment* (pp. 91–96). Alexandria, VA: American Statistical Association.

Griffith, D., Bennett, R., & Haining, R. (1989). Statistical analysis of spatial data in the presence of missing observations: A methodological guide and an application to urban census data. *Environment & Planning A, 18*, 1511–1523.

Kuhn, H., & Kuenne, R. (1962). An efficient algorithm for the numerical solution of the generalized Weber problem in spatial economics. *Journal of Regional Science, 4*, 21–34.

Ostresh, L. M. (1975). An efficient algorithm for solving the two center location-allocation problem. *Journal of Regional Science, 15*(2), 209–216.

Scott, A. (1970). Location-allocation systems: A review. *Geographical Analysis, 2*, 95–119.

Weber, A. (1909). *Über den Standort der Industrien.* Tubingen (Translated by C. Friedrich as *Alfred Weber's Theory of the Location of Industries*, 1929). Chicago: University of Chicago Press.

Chapter 3
Space–Time Autocorrelation

3.1 Introduction

In 1981, Cliff and Ord, in one development, and Griffith, in another, independently formulated a space–time version of the Moran Coefficient (MC), say MC_{ST}, as an extension of the static MC. The morphism of interest here is between MC and MC_{ST}. Chapter 14 addresses space–time issues in spatial econometrics. This current chapter explores spatial statistical properties associated with the two common space–time weights matrix specifications for MC_{ST}, namely the contemporaneous and the spatial lag structures. It also derives the asymptotic variance of MC_{ST} and explores some of its properties. Finally, eigenvector space–time filtering is reviewed, following Griffith (2012), but using the 2006–2013 employment classified into five general categories and geographically distributed across Poland by powiaty (NUTS4 ≈county; $n = 380$).

3.2 Specifying a Space–Time Moran Coefficient

The three-dimensional space–time matrix representation for the contemporaneous specification is given by

$$\mathbf{C}_{\mathrm{ST}} = \mathbf{I}_{\mathrm{T}} \otimes \mathbf{C}_{\mathrm{s}} + \mathbf{C}_{\mathrm{T}} \otimes \mathbf{I}_{\mathrm{s}}. \tag{3.1}$$

where $\otimes$ denotes Kronecker product, $\mathbf{I}_{\mathrm{T}}$ denotes the T-by-T identify matrix, $\mathbf{I}_{\mathrm{s}}$ denotes the n-by-n identity matrix, $\mathbf{C}_{\mathrm{T}}$ denotes the T-by-T time series connectivity matrix (i.e., the upper and lower off-diagonal cells contain 1, and all other cells contain 0), and $\mathbf{C}_{\mathrm{s}}$ denotes the n-by-n spatial weights matrix (defined here using the rook's adjacency criterion). Meanwhile, the three-dimensional connectivity matrix representation for the lagged specification is given by

D. A. Griffith, J. H. P. Paelinck, *Morphisms for Quantitative Spatial Analysis*, Advanced Studies in Theoretical and Applied Econometrics 51,
https://doi.org/10.1007/978-3-319-72553-6_3

$$\mathbf{C}_{\mathrm{ST}} = \mathbf{C}_{\mathrm{T}} \otimes \mathbf{C}_{\mathrm{s}} + \mathbf{C}_{\mathrm{T}} \otimes \mathbf{I}_{\mathrm{s}}, \tag{3.2}$$

Matrix $\mathbf{C}_{\mathrm{ST}}$ has dimensions nT-by-nT. An assumption here is that matrix $\mathbf{C}_{\mathrm{s}}$ does not change through time.

The eigenvalues of matrix $\mathbf{C}_{\mathrm{T}}$ are known (see Griffith 1996a): $2\cos\left(\frac{t\,\pi}{T+1}\right)$, $t = 1, 2, \ldots, T$. The eigenvalues of matrix $\mathbf{C}_{\mathrm{s}}$ are known for a P-by-Q regular square tessellation forming a complete rectangular region. In most cases, these latter eigenvalues must be calculated numerically or approximated.

$\mathrm{MC}_{\mathrm{ST}}$ may be written, using matrix notation and Eq. (3.1), as

$$\mathrm{MC}_{\mathrm{ST},1} = \frac{nT}{\mathbf{1}^{\mathrm{T}}\mathbf{C}_{\mathrm{ST}}\mathbf{1}} \frac{\mathbf{Y}^{\mathrm{T}}\left[\mathbf{I} - \mathbf{11}^{\mathrm{T}}/(nT)\right] \mathbf{C}_{\mathrm{ST}} \left[\mathbf{I} - \mathbf{11}^{\mathrm{T}}/(nT)\right] \mathbf{Y}}{\mathbf{Y}^{\mathrm{T}}\left[\mathbf{I} - \mathbf{11}^{\mathrm{T}}/(nT)\right] \mathbf{Y}}, \tag{3.3}$$

where nT-by-1 vector $\mathbf{Y}$ is a concatenation of T geographic distributions, one for each point in time, with the time index in descending order; $\mathbf{1}$ is an nT-by-1 vector of ones; and superscript T denotes the matrix transpose operation. The expected value of the $\mathrm{MC}_{\mathrm{ST}}$ as defined by Eq. (3.3) in the presence of zero spatial autocorrelation is a straightforward generalization of that for MC: $-1/(nT-1)$. This result may be derived as follows, positing an assumption of independent and identically distributed (iid) normal random variables (RVs):

$$\begin{aligned}
&E\left\{\frac{nT}{\mathbf{1}^{\mathrm{T}}\mathbf{C}_{\mathrm{ST}}\mathbf{1}} \frac{\mathbf{Y}^{\mathrm{T}}\left[\mathbf{I} - \mathbf{11}^{\mathrm{T}}/(nT)\right] \mathbf{C}_{\mathrm{ST}} \left[\mathbf{I} - \mathbf{11}^{\mathrm{T}}/(nT)\right] \mathbf{Y}}{\mathbf{Y}^{\mathrm{T}}\left[\mathbf{I} - \mathbf{11}^{\mathrm{T}}/(nT)\right] \mathbf{Y}}\right\} \\
&= \frac{nT}{\mathbf{1}^{\mathrm{T}}\mathbf{C}_{\mathrm{ST}}\mathbf{1}} E\left\{\frac{\mathbf{Y}^{\mathrm{T}}\left[\mathbf{I} - \mathbf{11}^{\mathrm{T}}/(nT)\right] \mathbf{C}_{\mathrm{ST}} \left[\mathbf{I} - \mathbf{11}^{\mathrm{T}}/(nT)\right] \mathbf{Y}}{\mathbf{Y}^{\mathrm{T}}\left[\mathbf{I} - \mathbf{11}^{\mathrm{T}}/(nT)\right] \mathbf{Y}}\right\} \\
&= \frac{nT}{\mathbf{1}^{\mathrm{T}}\mathbf{C}_{\mathrm{ST}}\mathbf{1}} \frac{E\left\{\mathbf{Y}^{\mathrm{T}}\left[\mathbf{I} - \mathbf{11}^{\mathrm{T}}/(nT)\right] \mathbf{C}_{\mathrm{ST}} \left[\mathbf{I} - \mathbf{11}^{\mathrm{T}}/(nT)\right] \mathbf{Y}\right\}}{E\left\{\mathbf{Y}^{\mathrm{T}}\left[\mathbf{I} - \mathbf{11}^{\mathrm{T}}/(nT)\right] \mathbf{Y}\right\}} \\
&= \frac{nT\left[-\sigma^2/(nT)\right]}{(nT-1)\,\sigma^2} = -1/(nT-1).
\end{aligned}$$

The same result is obtained with the randomization perspective.

$\mathrm{MC}_{\mathrm{ST}}$ may be written, using matrix notation and Eq. (3.2), as

$$\mathrm{MC}_{\mathrm{ST},2} = \frac{n(T-1)}{\mathbf{1}^{\mathrm{T}}\mathbf{C}_{\mathrm{ST}}\mathbf{1}} \frac{\mathbf{Y}^{\mathrm{T}}\left[\mathbf{I} - \mathbf{11}^{\mathrm{T}}/(nT)\right] \mathbf{C}_{\mathrm{ST}} \left[\mathbf{I} - \mathbf{11}^{\mathrm{T}}/(nT)\right] \mathbf{Y}}{\mathbf{Y}^{\mathrm{T}}\left[\mathbf{I} - \mathbf{11}^{\mathrm{T}}/(nT)\right] \mathbf{Y}}. \tag{3.4}$$

This is the specification appearing in Griffith (1981). Equation (3.1) has $(T-2)\mathbf{1}^{\mathrm{T}}\mathbf{C}_{\mathrm{s}}\mathbf{1}$ more links than Eq. (3.2); each of the two end points has only a single set of associations. If $T = 1$, Eq. (3.1) reduces to the static MC, whereas Eq. (3.2) vanishes. In other words, Eq. (3.1) is consistent with the conventional MC index, and Eq. (3.2) requires data for at least two points in time. Positing an assumption of iid normal RVs yields

$$
\begin{aligned}
&E\left\{\frac{n(T-1)}{\mathbf{1}^{\mathrm{T}}\mathbf{C}_{\mathrm{ST}}\mathbf{1}}\frac{\mathbf{Y}^{\mathrm{T}}\left[\mathbf{I}-\mathbf{1}\mathbf{1}^{\mathrm{T}}/(nT)\right]\mathbf{C}_{\mathrm{ST}}\left[\mathbf{I}-\mathbf{1}\mathbf{1}^{\mathrm{T}}/(nT)\right]\mathbf{Y}}{\mathbf{Y}^{\mathrm{T}}\left[\mathbf{I}-\mathbf{1}\mathbf{1}^{\mathrm{T}}/(nT)\right]\mathbf{Y}}\right\}\\
&=\frac{n(T-1)}{\mathbf{1}^{\mathrm{T}}\mathbf{C}_{\mathrm{ST}}\mathbf{1}}E\left\{\frac{\mathbf{Y}^{\mathrm{T}}\left[\mathbf{I}-\mathbf{1}\mathbf{1}^{\mathrm{T}}/(nT)\right]\mathbf{C}_{\mathrm{ST}}\left[\mathbf{I}-\mathbf{1}\mathbf{1}^{\mathrm{T}}/(nT)\right]\mathbf{Y}}{\mathbf{Y}^{\mathrm{T}}\left[\mathbf{I}-\mathbf{1}\mathbf{1}^{\mathrm{T}}/(nT)\right]\mathbf{Y}}\right\}\\
&=\frac{n(T-1)}{\mathbf{1}^{\mathrm{T}}\mathbf{C}_{\mathrm{ST}}\mathbf{1}}\frac{E\left\{\mathbf{Y}^{\mathrm{T}}\left[\mathbf{I}-\mathbf{1}\mathbf{1}^{\mathrm{T}}/(nT)\right]\mathbf{C}_{\mathrm{ST}}\left[\mathbf{I}-\mathbf{1}\mathbf{1}^{\mathrm{T}}/(nT)\right]\mathbf{Y}\right\}}{E\left\{\mathbf{Y}^{\mathrm{T}}\left[\mathbf{I}-\mathbf{1}\mathbf{1}^{\mathrm{T}}/(nT)\right]\mathbf{Y}\right\}}\\
&=\frac{n(T-1)\left[-\sigma^2/(nT)\right]}{(nT-1)\,\sigma^2}\\
&=-\frac{(T-1)}{T(nT-1)}.
\end{aligned}
$$

The same result is obtained with the randomization perspective. The ratio of these two expected values is

$$
\frac{E(\mathrm{MC}_{\mathrm{ST},1})}{E(\mathrm{MC}_{\mathrm{ST},2})}=-\frac{\frac{1}{nT-1}}{-\frac{(T-1)}{T(nT-1)}}=\frac{T}{T-1},
$$

which converges on 1 as T goes to infinity (i.e., in the limit, the two expected values become the same, namely 0).

The variance also differs between these two space–time connectivity matrix specifications. For Eq. (3.1), the asymptotic standard error is

$$
\sqrt{\frac{1}{T\mathbf{1}^{\mathrm{T}}\mathbf{C}_s\mathbf{1}/2+(T-1)n}}, \tag{3.5}
$$

whereas for Eq. (3.2), the asymptotic standard error is

$$
\sqrt{\frac{1}{(T-1)\left(\mathbf{1}^{\mathrm{T}}\mathbf{C}_s\mathbf{1}+n\right)}}. \tag{3.6}
$$

Replacing the normality assumption with one of randomization yields these same results.

3.3 Properties of the Space–Time Moran Coefficient

Figure 3.1 portrays the expected values for the two space–time connectivity matrices [i.e., Eqs. (3.1) and (3.2)]. Figure 3.1a portrays the expected value of Eq. (3.3) [i.e., $E(\mathrm{MC}_{\mathrm{ST},\,1})$] in green, with the expected value of Eq. (3.4) [i.e., $E(\mathrm{MC}_{\mathrm{ST},\,2})$]

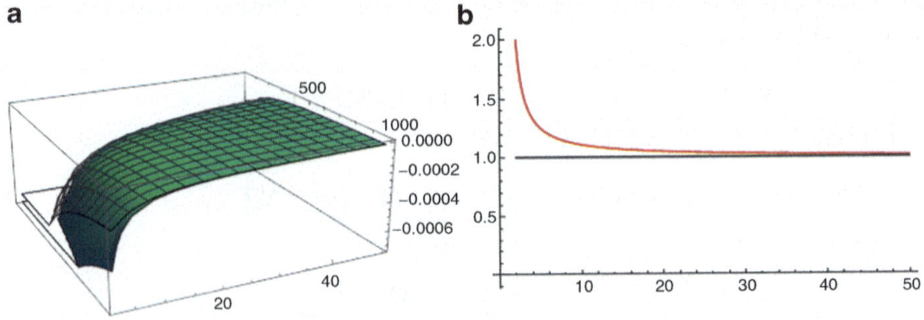

Fig. 3.1 Left (**a**): a plot of the expected value of Eq. (3.3) [i.e., $E(\mathrm{MC}_{\mathrm{ST},\,1})$] in green, with the expected value of Eq. (3.4) [i.e., $E(\mathrm{MC}_{\mathrm{ST},\,2})$] superimposed as a transparent mesh. Right (**b**): a plot of the ratio $E(\mathrm{MC}_{\mathrm{ST},\,1})/E(\mathrm{MC}_{\mathrm{ST},\,2})$, denoted by a red line, with the gray line denoting a ratio of 1

superimposed as a transparent mesh. Convergence of the latter is from below. Figure 3.1b shows the plot of the ratio of the two expected values [i.e., $E(\mathrm{MC}_{\mathrm{ST},\,1})/E(\mathrm{MC}_{\mathrm{ST},\,2})$], indicating that the difference between their expected values is inconsequential by $T = 50$.

The ratio of expressions (3.5) and (3.6) furnishes a measure of statistical efficiency. The four planar surface partitionings appearing in Fig. 3.2 can be used to exemplify this measure. Their asymptotic variance ratios [i.e., the square of expression (3.5) divided by the square of expression (3.6)] are as follows: 2/3 for linear, 3/5 for square, and 4/7 for hexagonal and maximum connectivity. Consequently, the lagged weights matrix specification yields a smaller standard error. Furthermore, the convergence mostly is determined by T, with little variation across n. The curves for the various surface partitionings, n, and T are concave upward and are very similar, but in some cases with different asymptotes. This feature can be seen by inspecting Fig. 3.3, which portrays the two extreme cases for $n = 30$, 100, and 1000, and $T = 2, 3, \ldots, 50$.

Besides expected values and variances, a $\mathrm{MC}_{\mathrm{ST}}$ index needs to be evaluated with regard to mixtures of RVs. The three most common RVs in spatial statistics and spatial econometrics are the normal, Poisson, and binomial (with the Bernoulli as a special case). The type of mixture one could anticipate in a space–time series would be the same RV for each point in time, but with parameters that vary across T. Table 3.1 summarizes results of a simulation experiment assessing Eqs. (3.3) and (3.4). The first part of the table demonstrates that the expected values are correct, and that expressions (3.5) and (3.6) furnish extremely accurate standard error estimates; none of the simulation results are significantly different from their theoretical counterparts. The second part summarizes results for illustrative parametric mixture models; these findings extend those reported in Griffith (2010). The specimen normal–normal mixture has the mean of the normal RV distributed as $\mathrm{N}(0, 0.1T)$; this is in keeping with the notion that the RV is constant, whereas its variance changes over time. Here, the space–time dataset contains variance heterogeneity. The specimen Poisson-gamma mixture (which relates to the negative binomial RV) has the mean of the Poisson RV distributed as a gamma RV with $\alpha = 10/T$ and $\beta = T/2$ to preserve a mean of 5. Here, the extra-Poisson variation

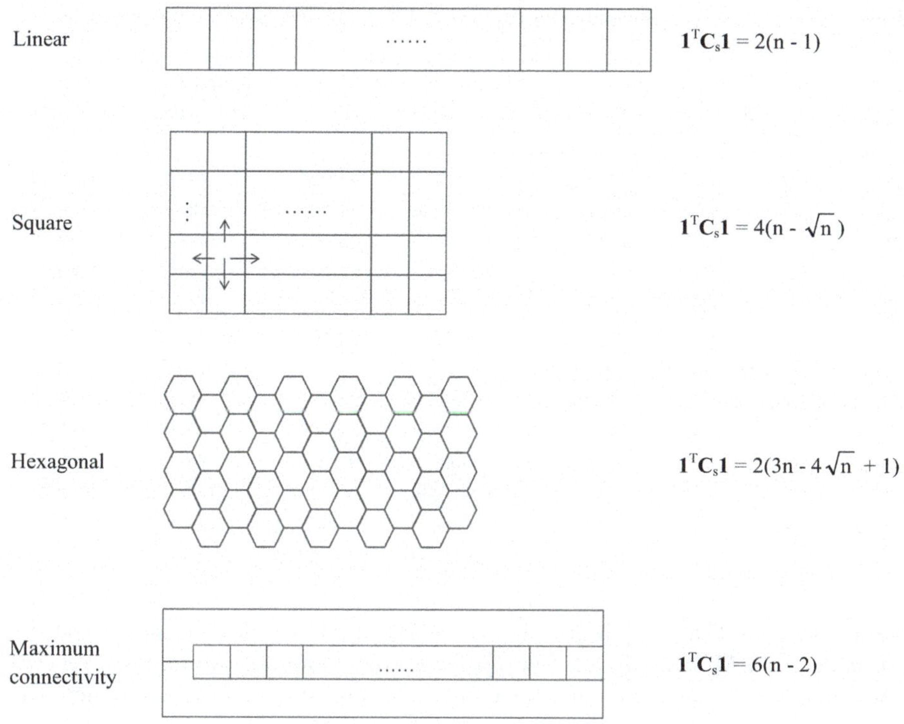

Fig. 3.2 Specimen geographic landscapes used to assess asymptotic statistical efficiency

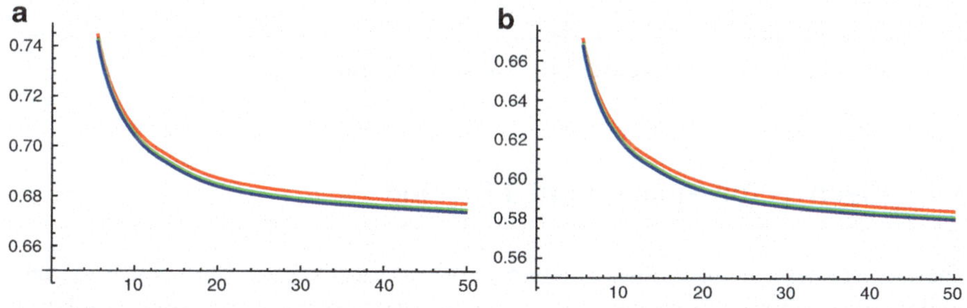

Fig. 3.3 Statistical efficiency plots for increasing T; red curves denote $n = 30$, green curves denote $n = 100$, and blue curves denote $n = 1000$. Left (**a**): linear surface partitioning. Right (**b**) maximum connectivity surface partitioning

increases over time, resulting in overdispersion characterizing the resulting counts. Finally, the specimen beta-binomial mixture has the mean of the binomial RV distributed as a beta RV with $\alpha = \beta = 0.1T$ to preserve ½ as the probability of an event occurring. Here, the extra binomial variation increases over time resulting in overdispersion characterizing the resulting percentages. The simulation results

Table 3.1 Analytical and simulation results for two MC_{ST} specifications, 7500 replications

		Contemporaneous connectivity matrix $\mathbf{C}_{ST}$		Lagged connectivity matrix $\mathbf{C}_{ST}$	
Random variable	Parameter	Theoretical	Simulation	Theoretical	Simulation
Independent and identically distributed					
Normal, N(0, 1)	$E(MC_{ST})$	−0.000376	−0.000434	−0.000322	−0.000249
	$\sigma_{MC_{ST}}$	0.010396	0.010234	0.008358	0.008383
Poisson, $\mu = 5$	$E(MC_{ST})$	−0.000376	−0.000468	−0.000322	−0.000202
	$\sigma_{MC_{ST}}$	0.010396	0.010242	0.008358	0.008367
Binomial, $p = 0.5$, $n_T = 1$	$E(MC_{ST})$	−0.000376	−0.000277	−0.000322	−0.000576
	$\sigma_{MC_{ST}}$	0.010396	0.010501	0.008358	0.008374
Independent mixtures of the same distribution					
Normal-normal, N(0, 1)	$E(MC_{ST})$	−0.000376	0.098736	−0.000322	−0.020747
	$\sigma_{MC_{ST}}$	0.010396	0.060405	0.008358	0.059297
Poisson-gamma, $\mu = 5$	$E(MC_{ST})$	−0.000376	0.394287	−0.000322	−0.087197
	$\sigma_{MC_{ST}}$	0.010396	0.117642	0.008358	0.195922
Beta-binomial, $p = 0.5$, $n_T = 1$	$E(MC_{ST})$	−0.000376	0.386572	−0.000322	−0.087881
	$\sigma_{MC_{ST}}$	0.010396	0.115793	0.008358	0.210634

summarized in Table 3.1 imply that these RV mixtures bias the mean and the variance of the MC_{ST}, with less bias for the normal–normal mixture, less expected value bias for the lagged connectivity matrix specification, and less variance bias for the contemporaneous connectivity matrix specification. Nevertheless, all bias is statistically significant.

Extending findings by de Jong et al. (1984), the maximum eigenvalue for connectivity matrix $[\mathbf{I} - \mathbf{11}^T/(nT)]\mathbf{C}_{ST}[\mathbf{I} - \mathbf{11}^T/(nT)]$ is 1.151166 for the contemporaneous case and 1.236338 for the lagged case. Accordingly, the mixture model biases are a bit less than suggested by Table 3.1 results.

3.4 Eigenvector Space–Time Filtering

The numerators[1] of the Polish location quotients (LQNs) are binomial RVs, and contain excess binomial variation from at least two sources, spatial autocorrelation and heterogeneity. This RV always remains the same across time, but with different parameters. The first data assessment is to quantify the degree of overdispersion across years. Table 3.2 summarizes the multiplicative factor for the various LQNs by year, and for the pooled years. The range of overdispersion is not excessive across the years for each employment category. Each employment category index

[1]These binomial RVs have been standardized to a denominator of 100 to control for variation due to local labor market size. Furthermore, these RVs need to be divided by the national percentage in their corresponding employment category to convert them to LQs.

Table 3.2 Extra-binomial variation parameter estimate for annual Polish LQNs, 2007–2013

Year	$LQN_{agriculture}$	$LQN_{industry}$	LQN_{trade}	$LQN_{finance}$	$LQN_{services}$
2007	**4.6892**	**2.7065**	**2.0086**	1.1361	1.8014
2008	4.6420	2.6730	1.9896	**1.1056**	**1.7794**
2009	4.6475	2.6295	1.9873	1.1404	1.8260
2010	4.4917	2.5909	1.9530	1.1643	1.8301
2011	4.4916	2.5889	1.9292	1.1847	1.8488
2012	4.4962	2.6022	1.9273	**1.1850**	**1.8967**
2013	**4.4661**	**2.5686**	**1.9022**	1.1811	1.8941
Pooled	4.5585	2.6283	1.9580	1.1568	1.8392
Yearly extremes ratio	1.0500	1.0537	1.0559	1.0718	1.0659

Note: Bold denotes an annual extreme

also displays an apparent time series trend (although seven points in time are insufficient to determine a true trend). $LQN_{agriculture}$ exhibits sizeable overdispersion; $LQN_{finance}$ exhibits nearly trivial overdispersion.

Griffith (2012) presents the Moran eigenvector space–time filtering methodology, which extends its static spatial filtering counterpart (Griffith 2003). In this context, eigenvector space–time filtering uses a set of space–time proxy variables, which are extracted as eigenvectors from an nT-by-nT modified (i.e., doubly centered) binary space–time relationship matrix, $\mathbf{C}_{ST}$—either Eq. (3.1) or (3.2), which has 1s in the cells whose row and column locations are neighbors in space and/or time, and 0s elsewhere—that ties geographic observations together, and adds these vectors as control variables into a model specification. Griffith (1996b) describes eigenfunctions for these matrices. Paralleling the static methodology, these control variables identify and isolate stochastic space–time dependencies among the location-and-time-indexed observations, thus allowing model building to proceed with observations that mimic independent ones.

Similar to the space-only case, because substituting the eigenvectors into the matrix expression $[\mathbf{I} - \mathbf{11}^T/(nT)]\mathbf{C}_{ST}[\mathbf{I} - \mathbf{11}^T/(nT)]$ results in a Rayleigh quotient, with vector $\mathbf{E}_1$ maximizing this expression; these eigenvectors can be interpreted as follows:

the first eigenvector, say $\mathbf{E}_1$, is the set of real numbers that has the largest space–time MC_{ST} achievable by any set for the areal unit articulation defined by the space–time connectivity matrix $\mathbf{C}_{ST}$; the second eigenvector is the set of real numbers that has the largest achievable space–time MC_{ST} by any set that is orthogonal and uncorrelated with $\mathbf{E}_1$; the third eigenvector is the third such set of real numbers; and so on through $\mathbf{E}_n$, the set of real numbers that has the largest negative space–time MC_{ST} achievable by any set that is orthogonal and uncorrelated with the preceding $(n-1)$ eigenvectors.

Accordingly, these eigenvectors furnish distinct map patterns that most likely change through time in some structured way (i.e., the Kronecker products are time-specific values that multiply geographic map pattern values) and hence furnish descriptions of latent space–time autocorrelation in a response variable. An

Table 3.3 ESTF summary results

Summary statistic	$LQN_{agriculture}$	$LQN_{industry}$	LQN_{trade}	$LQN_{finance}$	$LQN_{services}$
Contemporaneous $\mathbf{C}_{ST}$					
# eigenvectors	160	130	155	136	138
Pseudo-R^2	0.7761	0.6404	0.6906	0.6411	0.5727
a	−0.0002	0.0007	0.0006	−0.0008	−0.0015
b	1.0008	0.9977	0.9957	1.0316	1.0059
Over-dispersion parameter	2.5128	1.6503	1.1095	0.6538	1.2313
% of Table 3.2 pooled	55.13	62.79	56.67	56.52	66.95
Lagged $\mathbf{C}_{ST}$					
# eigenvectors	91	95	96	84	83
Pseudo-R^2	0.6565	0.5762	0.5031	0.4642	0.3769
a	−0.0079	0.0017	0.0018	−0.0008	0.0014
b	1.0265	0.9940	0.9872	1.0320	0.9944
Over-dispersion parameter	3.1896	1.7663	1.3930	0.7898	1.4624
% of Table 3.2 pooled	69.97	67.20	71.15	68.28	79.51

Note: The candidate set from contemporaneous matrix $\mathbf{C}_{ST}$ comprises 573 eigenvectors; the candidate set from lagged matrix $\mathbf{C}_{ST}$ comprises 264 eigenvectors

eigenvector space–time filter (ESTF) may be constructed as a linear combination of judiciously selected vectors from some candidate subset of these eigenvectors. A threshold value of $MC_{ST}/MAX(MC_{ST}) = 0.25$ is used here to demarcate the candidate subsets, which consist of 573 prominent eigenvectors for the contemporaneous specification [i.e., Eq. (3.1)] and 264 prominent eigenvectors for the lagged specification [i.e., Eq. (3.2)]; the total number of eigenvectors is 2660.

Summary results appearing in Table 3.3 suggest that the contemporaneous outperforms the lagged $\mathbf{C}_{ST}$ specification. In other words, current geographic neighbor LQs coupled with in situ LQs from a preceding time period account for much of the space–time variation in LQs across Poland by powiat. Of note is that this dataset contains a single employment category exhibiting underdispersion, an infrequently encountered phenomenon.

3.5 Omitted Variables in a Description of Space–Time Response Variables

An applied spatial scientist spends considerable time deciding about which covariates, as well as their functional forms, to include in a descriptive regression equation. Misspecification of this descriptive equation by failing to include all relevant covariates (i.e., the omitted variables problem), or including superfluous variables, is a serious challenge in spatial regression analysis. The Ramsey (1969) regression equation specification error test (RESET) furnishes a technique to assess impacts of omitted variables. Its assessment is indirect; it addresses whether or not

Table 3.4 RESET statistics for a GLM description of $LQN_{agriculture}$

		Unadjusted saturated specification		ESTF-adjusted saturated specification	
RESET term	df	Chi-square	Probability	Chi-square	Probability
$\hat{y}^2$	1	22.64	1.9×10^{-6}	3.22	0.0112
$\hat{y}^3$	2	29.92	3.2×10^{-7}	3.87	0.0209
$\hat{y}^4$	3	36.08	7.2×10^{-8}	3.87	0.0517

Note: The chi-square statistics is -2(log-likelihood$_{reduced}$ $-$ (log-likelihood$_{full}$)
df degrees of freedom

nonlinear combinations of fitted values help explain a response variable. Its supporting logic contends that these nonlinear combinations (e.g., exponential powers and cross-products) of covariates that correlate with a response variable signify a misspecified equation, implying the existence of omitted variables. Sapra (2005) extends Ramsey's initial linear regression-based technique to generalized linear models (GLMs).

$LQN_{agriculture}$ is used here for illustrative purposes. This illustration employs an additive approximately (not fully) saturated model specification with 6 binary 0-1 indicator variables denoting time periods and 379 binary 0-1 indicator variables denoting powiaty; this specification is not fully saturated because it does not contain indicator variable interaction terms. These indicator variables represent fixed effects that are point-in-time and location-in-space specific and account for heterogeneity (e.g., each time period has its own mean, and each powiat has its own mean) that otherwise would be accounted for by omitted variables. In other words, they are surrogates for unknown substantive variables. In addition, the specification includes both density and percentage of working age population. The ESTF is based upon Eq. (3.1) and is a linear combination of 41 of the 573 candidate eigenvectors.

Findings summarized in Table 3.4 imply that the ESTF compensates reasonably well for omitted variables not accounted for by the fixed effects terms. This outcome is attributable, in part, to the sharing of common eigenvectors by space–time RVs, which tends to increase detected multicollinearity amongst covariates.

3.6 Concluding Comments

This chapter, whose principal topic is space–time data analysis, revisits the literature of the early 1980s, focusing on the isomorphism between MC and MC_{ST}. It contributes selected new asymptotic results for MC_{ST}. It also contributes findings based upon a wide range of surface partitionings. Another of its contributions is the mixture distribution findings, which indicate that overdispersion can complicate MC_{ST} interpretations by introducing different biases into the calculation of this index. Yet another of its contributions is the comparative findings about the two different definitions of matrix $\mathbf{C}_{ST}$; the contemporaneous definition appears to be preferable to the lagged definition. Finally, the RESET analysis suggests that an

ESTF can do a very good job of compensating for omitted variables in a regression equation specification. This latter result is corroborated by the pseudo-R^2 values reported in Table 3.3: an ESTF tends to account for 50–75% of the geographic variation in LQs.

References

Cliff, A., & Ord, J. (1981). Spatial and temporal analysis: Autocorrelation in space and time. In N. Wrigley & R. Bennett (Eds.), *Quantitative geography: A British view* (pp. 104–110). London: Routledge & Kegan Paul.

de Jong, P., Sprenger, C., & van Veen, F. (1984). On extreme values of Moran's I and Geary's c. *Geographical Analysis, 16*(1), 17–24.

Griffith, D. (1981). Interdependence in space and time: Numerical and interpretative considerations. In D. Griffith & R. MacKinnon (Eds.), *Dynamic spatial models* (pp. 258–287). Alphen aan den Rijn: Sijthoff and Noordhoff.

Griffith, D. (1996a). Spatial statistical analysis and GIS: Exploring computational simplifications for estimating the neighborhood spatial forecasting model. In P. Longley & M. Batty (Eds.), *Spatial analysis: Modelling in a GIS environment* (pp. 255–268). Haarlow: Longman GeoInformation.

Griffith, D. (1996b). Spatial autocorrelation and eigenfunctions of the geographic weights matrix accompanying geo-referenced data. *The Canadian Geographer, 40*, 351–367.

Griffith, D. (2003). *Spatial autocorrelation and spatial filtering: Gaining understanding through theory and scientific visualization*. Berlin: Springer.

Griffith, D. (2010). The Moran Coefficient for non-normal data. *Journal of Statistical Planning and Inference, 140*, 2980–2990.

Griffith, D. (2012). Space, time, and space-time eigenvector filter specifications that account for autocorrelation. *Estadística Española, 54*(#177), 7–34.

Ramsey, J. (1969). Tests for specification errors in classical linear least squares regression analysis. *Journal of the Royal Statistical Society Series B, 31, 350–371.*

Sapra, S. (2005). A regression error specification test (RESET) for generalized linear model. *Economics Bulletin, 3*(1), 1–6.

Chapter 4
The Relative Importance of Spatial and Temporal Autocorrelation

4.1 Introduction

One clear distinction between spatial and temporal correlation is that spatial is two-directional and two-dimensional, whereas temporal is one-directional and one-dimensional. One outcome of this structure is that temporal correlation tends to dominate space–time data dependencies. Algebraically, this dominance can be demonstrated with the calculus of expectations, where ρ denotes the autocorrelation parameter:

$$E(Y_t) = \rho E(Y_{t-1} + \epsilon_t)$$
$$\Rightarrow \mu = \rho\mu$$

Therefore, the value observed at time t is a partial amount of the value existing at time $t-1$ (i.e., inertia in the existence of a phenomenon). If ρ~beta[16, 16] $\approx$ N $(0.5, 0.087^2)$, then the probability $P(\rho \geq 0.9) = 7 \times 10^{-9}$. Shifting the mean of the beta distribution to 0.9, if ρ~beta[120, 13] such that $\rho^{6.6} \approx$ N$(0.52, 0.093^2)$, then the probability $P(\rho \geq 0.9) = 0.57$. If ρ~U[0, 1], then the probability $P(\rho \geq 0.9) = 0.10$. In contrast, for a spatial series geographically distributed across a regular square tessellation with a dependency structure based upon a rook's definition of adjacency and assuming that the spatial autocorrelation (SA) parameters ρ_j ($j = 1, 2, 3, 4$) are independent,

$$E(Y_{i,j}) = E\left[\rho_1 Y_{i-1,j} + \rho_2 Y_{i+1,j} + \rho_3 Y_{i,j-1} + \rho_4 Y_{i,j+1}\right)/4 + \epsilon_t\right]$$
$$\Rightarrow \mu = \frac{\sum_{j=1}^{4} \rho_j}{4}\mu = \rho\mu$$

Therefore, the value observed at location (i,j) is a partial amount of the values existing at its four neighboring locations (i.e., similar values of a phenomenon tend

D. A. Griffith, J. H. P. Paelinck, *Morphisms for Quantitative Spatial Analysis*, Advanced Studies in Theoretical and Applied Econometrics 51,
https://doi.org/10.1007/978-3-319-72553-6_4

Table 4.1 Temporal correlation with 2013 LQs

	Year					
Employment category	2012	2011	2010	2009	2008	2007
A	0.99945	0.99887	0.99836	0.97211	0.97085	0.96943
I	0.99416	0.98888	0.98514	0.96547	0.95815	0.95532
T	0.99333	0.98627	0.97580	0.96084	0.94293	0.92902
F	0.98199	0.97582	0.96503	0.95165	0.93373	0.92437
S	0.99262	0.98803	0.98494	0.95243	0.94375	0.92775

to agglomerate in space). If ρ_j~beta[16, 16] $\approx$ N$(0.5, 0.087^2)$, then the probability $P(\rho \geq 0.9) = 2 \times 10^{-33}$, which is markedly less than 7×10^{-9}. Shifting the mean of the beta distribution to 0.9, if ρ_j~beta[120, 13], then the probability $P(\rho \geq 0.9) = 0.10 \ll 0.57$. Finally, if ρ_j~U[0, 1], then the probability $P(\rho \geq 0.9) = 0.0001 \ll 0.10$. This demonstration illustrates why high levels of SA may tend to be more difficult to observe vis-à-vis the more commonly observed high levels of temporal autocorrelation.

The combinatorial nature of averaging multiple relationships for spatial data also increases the probability of observing intermediate values; the most commonly observed levels of SA for socioeconomic/demographic variables fall into the interval [0.4, 0.6]. For the preceding uniform distribution example, $P(0.4 \leq \rho \leq 0.6) = 0.497$ for the SA case, which is substantially greater than its 0.2 counterpart value for the time series case. Meanwhile, for the preceding beta[16, 16] distribution example, $P(0.4 \leq \rho \leq 0.6) = 0.979$ (an extremely accurate normal approximation) for the SA case, which is noticeably greater than its 0.743 counterpart value for the time series case.

Again the 2007–2013 employment by five general categories across Poland for powiaty (NUTS4 $\approx$county; $n = 380$) furnishes empirical examples for this chapter. Tables 4.1 and 4.2 illustrate the preceding discussions about correlation in space and in time: the time series of maps have very high positive correlation, with this correlation decreasing with increasing separation in time, and SA tends to be moderate [the adjusted Moran Coefficient (MC) is a statistically more powerful indicator of SA than the adjusted Geary Ratio (GR)]. Figure 4.1 portrays typical scatterplots of these relationships.

4.2 Random Effects: SSRE and SURE Components

Space–time data often are referred to as panel data (i.e., longitudinal or cross-sectional time-series data) because areal unit measurements are repeated across time. One perspective about such space–time data contends that model residuals

Table 4.2 Spatial autocorrelation (SA) indices for LQs

Employment category	SA index	Year						
		2013	2012	2011	2010	2009	2008	2007
A	MC*	0.45084	0.44987	0.44825	0.44868	0.45965	0.45430	0.45154
	GR	0.60037	0.59696	0.59241	0.59239	0.59017	0.58494	0.57854
I	MC*	0.38702	0.40338	0.40542	0.40147	0.41007	0.40346	0.39357
	GR	0.43924	0.45523	0.45139	0.44873	0.44894	0.44781	0.42966
T	MC*	0.32343	0.31225	0.30427	0.29921	0.30318	0.28621	0.27564
	GR	0.48170	0.48001	0.48689	0.48889	0.50607	0.49305	0.49110
F	MC*	0.21641	0.22473	0.22534	0.20863	0.21044	0.20253	0.20879
	GR*	0.44349	0.43125	0.42861	0.44558	0.43940	0.44220	0.44086
S	MC*	0.20684	0.20147	0.20506	0.20739	0.22853	0.21363	0.21252
	GR	0.46678	0.46204	0.45713	0.45022	0.46816	0.45385	0.44054

Note: $MC^* = MC/MC_{max}$ & $GR^* = 1-2GR/GR_{max}$; rescales positive SA indices to the interval (0, 1]

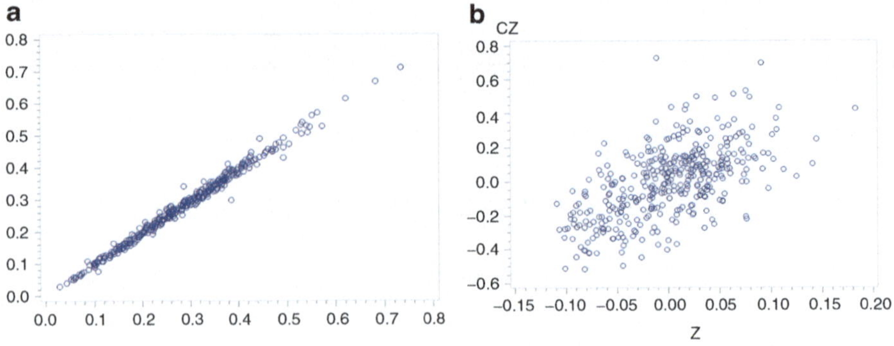

Fig. 4.1 Left (**a**): a typical time series scatterplot (2013 versus 2012 $LQ_{agriculture}$). Right (**b**): a typical Moran scatterplot (2013 $LQ_{agriculture}$)

contain two components, one that is the usual stochastic noise and another that is an individual areal unit specific effect.[1] This latter component may be labeled a random effect, can be cast as unobserved heterogeneity that is constant over time (i.e., an individual mean for each areal unit that differs from the global mean for all areal units), and signifies that the clustering of LQs in powiaty is a feature of interest rather than a data analysis nuisance. A critical model specification assumption is that random effects are uncorrelated with covariates in the specification. This assumption may fail to be satisfied because the set of areal units constituting a geographic landscape is a nonrandom sample.

Gelfand et al. (2003) employ this conceptualization as a mechanism to specify spatial models in which regression coefficients can vary at a local level rather than be globally constant across a given geographic landscape. Analyses in this chapter treat pure SA model specifications, deemphasizing the random effects assumption.

The general model specification may be written as follows:

$$\mathbf{Y} = \mathbf{X}\boldsymbol{\beta} + \boldsymbol{\xi} + \boldsymbol{\epsilon} \tag{4.1}$$

where $\mathbf{Y}$ is an nT-by-1 vector of space–time response variable data, $\mathbf{X}$ is an nT-by-(p+1) matrix of p covariates, $\boldsymbol{\beta}$ is a (p+1)-by-1 vector of regression coefficients, $\boldsymbol{\xi}$ is an n-by-1 vector of random effects that is concatenated T times, and $\boldsymbol{\epsilon}$ is an nT-by-1 vector of independent and identically distributed (iid) stochastic error. The random effects are constant across time: ξ_i is the same for all T. When $T = 1$, the residual is $\boldsymbol{\xi} + \boldsymbol{\epsilon}$.

The random effects term can be decomposed into two parts: a spatially structured (SSRE) and a spatially unstructured (SURE) part. Griffith (2008), Griffith and Paelinck (2009), and Griffith and Chun (2014) discuss this decomposition. The version of Eq. (4.1) utilized here is

[1] Separating these two components requires additional information, which repeated measures across time furnish. From a Bayesian viewpoint, prior distributions are an alternative for furnishing additional information.

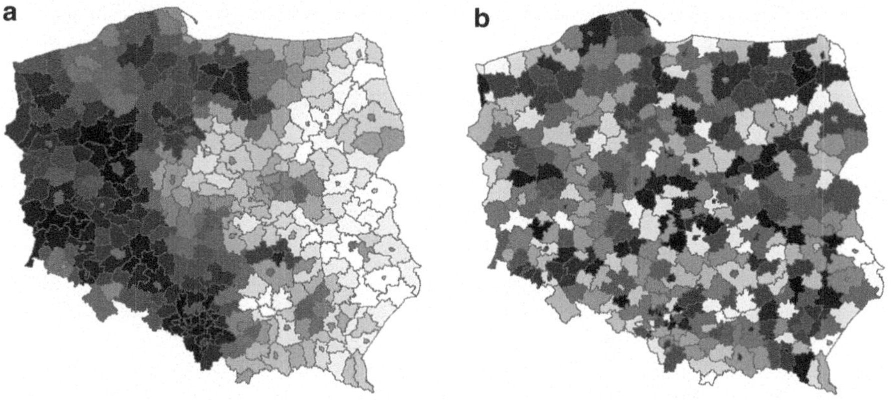

Fig. 4.2 The darkness of the grayscale is directly proportional to the magnitude of the values. Left (**a**): $LQ_{industry}$ SSRE. Right (**b**): $LQ_{industry}$ SURE

$$y_{it} = \mu + {}_{SSRE}\mu_i + {}_{SURE}\mu_i + \epsilon_{it} \quad (i = 1, 2, \ldots, 380; t = 1, 2, \ldots, 7) \tag{4.2}$$

where ${}_{SSRE}\mu_i$ captures SA and ${}_{SSRE}\mu_i + {}_{SURE}\mu_i$ is the time invariant areal unit i specific term that accounts for temporal correlation.

Consider the LQs for industrial employment (I). Figure 4.2 portrays the SSRE (Fig. 4.2a) and the SURE (Fig. 4.2b) terms. The composite random effects term fails to conform closely to a bell-shaped curve [P(Shapiro-Wilk) < 0.0001], whereas the SURE does [P(Shapiro-Wilk) $= 0.1786$]. The mean of the random effects term is 0.00192, which is nearly zero; the mean of the SSRE is exactly zero. The variability of the random effects term is nearly equally distributed across the SSRE and the SURE terms, with the SSRE term containing 28 eigenvectors and accounting for 53.30% of it. The MC for the SSRE term is 0.96046 (adjusted: 0.96046/1.17485 = 0.81752); the MC for the SURE term is −0.05872, indicating the presence of only trace SA.

One contention is that the SSRE can be estimated from a single point in time because it extracts the SA component of a random effects term. Results established by Patuelli et al. (2011) support this contention, revealing a set of common eigenvectors across years, combined with a mixture of infrequent eigenvectors by year. Table 4.3 corroborates this contention for the Polish $LQ_{industry}$; most eigenvectors in the ESF for the time-series-based SSRE also are common to individual year ESF. The SSRE accounts for roughly the same amount of variance in $LQ_{industry}$ for the repeated measures cases as well as for each of the individual years. Table 4.4 reports the correlations between the repeated measures and the individual annual ESFs by employment category.

The correlations between the space–time repeated measures SSRE term and the individual annual ESFs are very strong for all employment categories. This proof of concept analysis suggests that simulation experiments need to be executed to

Table 4.3 Eigenvector structure of annual vis-à-vis time-series-based SSRE estimates for $LQ_{industry}$

SSRE eigenvectors	Year 2007	2008	2009	2010	2011	2012	2013
E_1	1	1	1	1	1	1	1
E_2	1	1	1	1	1	1	1
E_6	1	1	1	1	1	1	1
E_8	1	1	1	1	1	1	1
E_9	1	1	1	1	1	1	1
E_{11}	1	1	1				
E_{13}	1	1	1	1	1	1	1
E_{15}	1	1	1	1	1	1	1
E_{18}	1	1	1	1	1	1	1
E_{19}	1	1	1	1	1	1	1
E_{22}	1	1	1	1	1	1	1
E_{25}	1	1	1	1	1	1	1
E_{26}							1
E_{31}							
E_{42}	1	1	1				
E_{43}							
E_{44}	1	1	1	1	1	1	1
E_{46}				1	1	1	1
E_{55}	1	1	1				
E_{58}	1	1	1	1	1	1	1
E_{59}	1	1	1	1	1	1	1
E_{65}	1	1	1	1	1	1	1
E_{68}				1	1	1	1
E_{70}	1	1	1	1	1	1	1
E_{75}	1	1	1	1	1	1	1
E_{78}	1	1	1				
E_{82}	1	1	1	1	1	1	1
E_{83}					1	1	
Others	2	2	2	3	3	1	1
R^2: 0.5330	0.50344	0.50985	0.51677	0.51073	0.52160	0.51335	0.49463

Table 4.4 Product moment correlation coefficients quantifying relationships between individual annual ESFs and repeated measures

Employment category	Year 2007	2008	2009	2010	2011	2012	2013
A	0.82717	0.82173	0.83477	0.85853	0.85652	0.85817	0.85242
I	0.95870	0.95853	0.95835	0.94968	0.95143	0.95907	0.96141
T	0.92210	0.90062	0.92247	0.93409	0.90298	0.90377	0.91146
F	0.84445	0.89909	0.90453	0.87254	0.85441	0.81685	0.80858
S	0.89901	0.90939	0.93479	0.83686	0.85968	0.89933	0.86283

confirm the relationship. If the SSRE term can be estimated without additional information beyond the eigenfunctions of a spatial weights matrix, then attention can be concentrated on how to estimate the SURE term without additional information. This finding is an important contribution of this book chapter.

4.3 Estimating a SURE Term: A Sensitivity Analysis

In the context of repeated measures, the preceding section establishes that a single time period allows the estimation of a SSRE term. Equation (4.2) indicates that what remains in a residual is a combination of a SURE term and a stochastic term.

Estimating a SURE term with a pair of consecutive time periods yields the correlations between these individual consecutive-pair-based results and the SUREs based upon the entire 7 years of data that appear in Table 4.5.

A tendency in these results is for the closest correspondences to be in the middle of the time series. These correlations suggest that the SURE term can be accurately estimated from only two consecutive points in time.

Table 4.6 reports correlations between the SURE terms based upon the entire 7 years of data and a pair of maps with increasing time between them:

Again, these correlations suggest that the SURE term can be accurately estimated from only two points in time, even when those time points are relatively far apart.

Rather than repeated measures, prior distributions in a Bayesian analysis framework furnish another source of additional information that could allow $t = 1$ SURE estimation. For the Polish LQs, the Bayesian analysis was executed with SAS PROC MCMC, which implements Markov chain Monte Carlo (MCMC) estimation. The burn-in number of iterations was 10,000, thinning was set at 200, and chains of length 210,000 were constructed; this yielded chains of length 1000 for estimation purposes (the numerical intensity of this exercise prevented the generation of longer chains). Normal priors, with means of 0 and variances of 100^2, were attached to both the SURE term and the coefficient for the estimated SSRE term that was part of the Bayesian analysis input. The prior for the variance of the individual SURE terms was a gamma distribution. The mixture time series plots

Table 4.5 Product moment correlations quantifying relationships between SUREs based upon pairs of consecutive time periods, and the entire 7-year time series

Employment category	Year					
	2007/2008	2008/2009	2009/2010	2010/2011	2011/2012	2012/2013
A	0.98397	0.98391	0.99743	0.99038	0.99028	0.98962
I	0.96260	0.97423	0.98298	0.97593	0.97045	0.96538
T	0.95124	0.96608	0.98218	0.98059	0.96080	0.96156
F	0.92870	0.93892	0.93827	0.94074	0.92199	0.93057
S	0.96003	0.96694	0.99504	0.97852	0.97136	0.96892

Table 4.6 Product moment correlations quantifying relationships between SUREs based upon pairs of consecutive time periods with increasing number of years between them and the entire 7-year time series

Employment category	Year					
	2007/2008	2007/2009	2007/2010	2007/2011	2007/2012	2007/2013
A	0.98397	0.98423	0.99409	0.99734	0.99456	0.99764
I	0.96260	0.97117	0.97646	0.98381	0.98818	0.98580
T	0.95124	0.96764	0.97524	0.96714	0.97031	0.97749
F	0.92870	0.94405	0.95206	0.95418	0.95257	0.94725
S	0.96003	0.96300	0.98595	0.98398	0.98438	0.98590

Table 4.7 Product moment correlations quantifying relationships between SUREs based upon a Bayesian analysis and the entire 7-year time series

Employment category	Year						
	2007	2008	2009	2010	2011	2012	2013
A	0.97790	0.97763	0.97837	0.98815	0.98842	0.98832	0.98738
I	0.96234	0.96809	0.97249	0.97971	0.97895	0.97488	0.96894
T	0.95056	0.96044	0.97330	0.97710	0.97622	0.97012	0.96193
F	0.87408	0.87955	0.88566	0.86021	0.87227	0.87100	0.86319
S	0.95349	0.97074	0.97601	0.98166	0.98185	0.98254	0.97747

and the time series correlograms display good diagnostic behavior ($LQ_{agriculture}$ exhibits the poorest performances). The medians of the posterior distributions were used to quantify the SUREs. Table 4.7 reports correlations between the SURE terms based upon the entire 7 years of data and those estimated with Bayesian techniques for $t = 1$:

Again, these correlations suggest that Bayesian techniques can produce an accurate estimate of the SURE term in the absence of repeated measures (i.e., $t = 1$).

Implications of findings summarized in this section include that a relatively small amount of data are needed to obtain good estimates of SSRE and SURE terms. This successful demonstration of concept needs to undergo more rigorous evaluation, including comprehensive simulation experiments involving lengthier space–time series. One appealing finding is that estimation techniques appear to exist that allow the SSRE term and its accompanying SURE term to be estimated when $t = 1$.

Table 4.8 Bivariate regression results for LQs regressed on imputed LQs: time beats space

Statistic	A	I	T	F	S
Temporal prediction					
Intercept	0.0003	0.0003	0.0001	0.0008	−0.0000
Slope	0.9988	0.9989	0.9995	0.9644	1.0003
Pseudo-R^2	0.9874	0.9757	0.9671	0.9467	0.9722
Space-only prediction					
Intercept	−0.0044	0.0033	0.0002	0.0001	0.0064
Slope	1.0145	0.9880	0.9989	0.9943	0.9743
Pseudo-R^2	0.4229	0.4944	0.4053	0.2669	0.22249

4.4 Time Beats Space

The random effects term, which comprises a SSRE and a SURE term, is invariant over time. In other words, once it is estimated, its value for a given areal unit is the same for each point in time. Griffith and Paelinck (2009) and Griffith and Chun (2014) show that this term can be used for short range space–time forecasting. As a forecasting horizon increases, such forecasts should become less reliable.

Table 4.1 highlights these two preceding time series features: the correlations across time are very strong (frequently much stronger than those across space; Table 4.2), and temporal correlations decrease in strength as the amount of intervening time increases. In other words, time beats space in terms of the quality of short range forecasts. Table 4.8 summarizes cross-validation results, where temporal prediction uses both the SURE and SSRE terms, and space-only prediction uses only the SSRE terms. In both cases, the trends are correct, with an intercept essentially equal to zero, and a slope essentially equal to one. Conspicuous differences include a closer correspondence between the temporal predicted and observed vis-à-vis space-only predicted and observed. Scatterplots portrayed in Fig. 4.3 accentuate this last point.

4.5 Space Beats Time

Inertia in a system coupled with unidirectional and one-dimensional dependency almost always enables temporal correlation to dominate spatial correlation. But if the invariant component (i.e., the SURE term) becomes scrambled (e.g., in 2005, Hurricane Katrina devastated New Orleans), the temporal correlation disappears, at least in the short run. Or, if sufficient time elapses between two points in time (see Griffith 2013), temporal correlation becomes weaker than spatial correlation. In these situations, spatial prediction outperforms temporal prediction.

Simulating this context, the SURE term was replaced by a random normal deviate with mean zero and variance 0.4^2 (the characteristics of the SURE term).

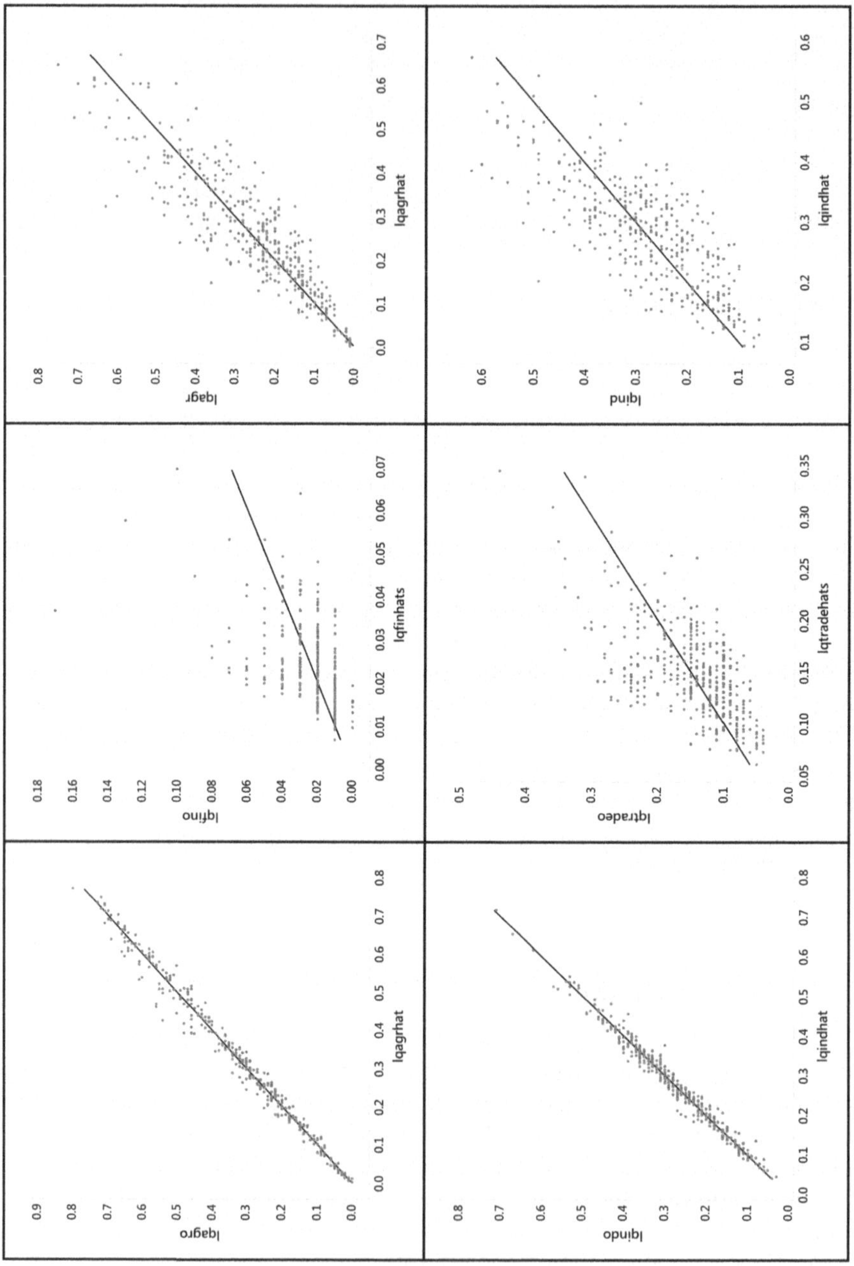

Fig. 4.3 When time beats space, the left-hand column portrays temporal prediction results, whereas the middle column portrays space-only prediction results; the right-hand column portrays when space beats time. The rows, respectively, are for the five employment categories: A, I, T, F, and S

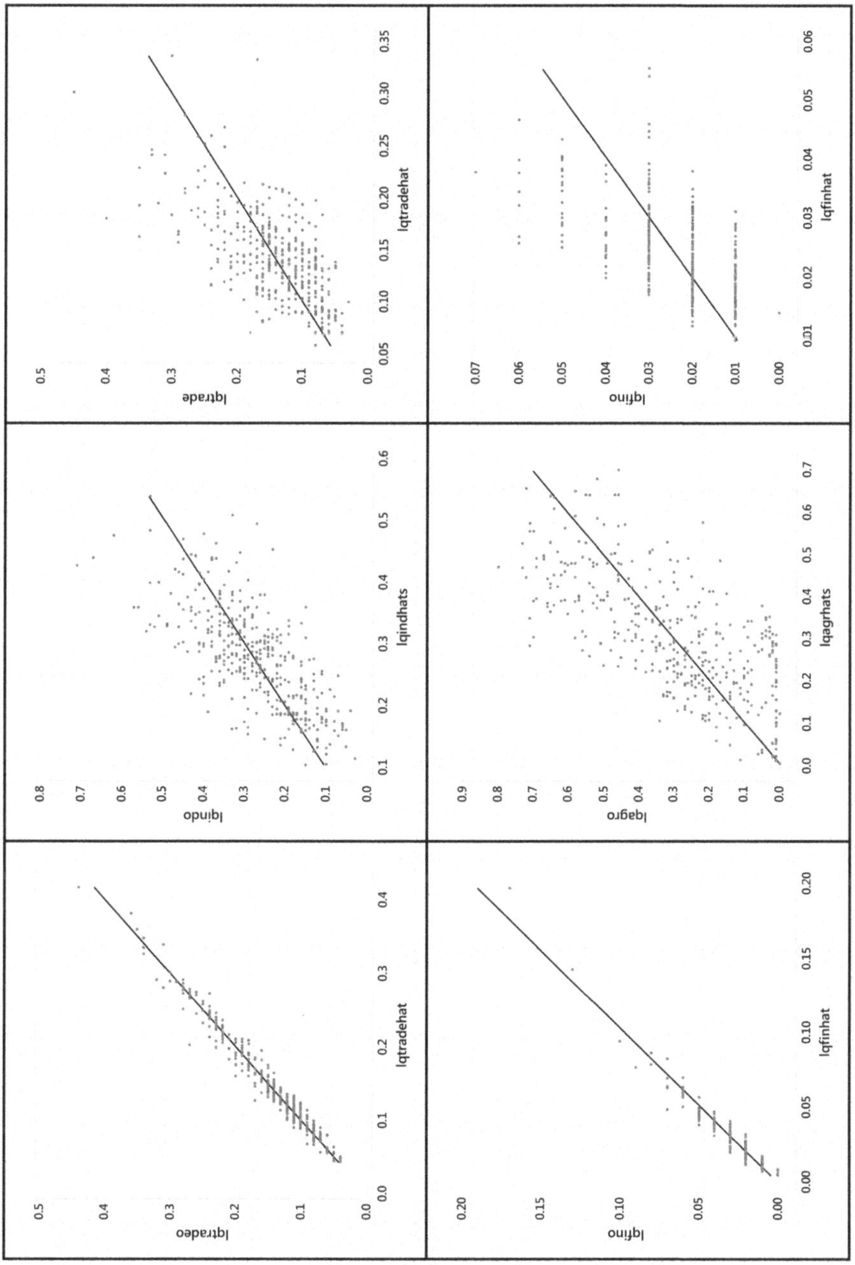

Fig. 4.3 (continued)

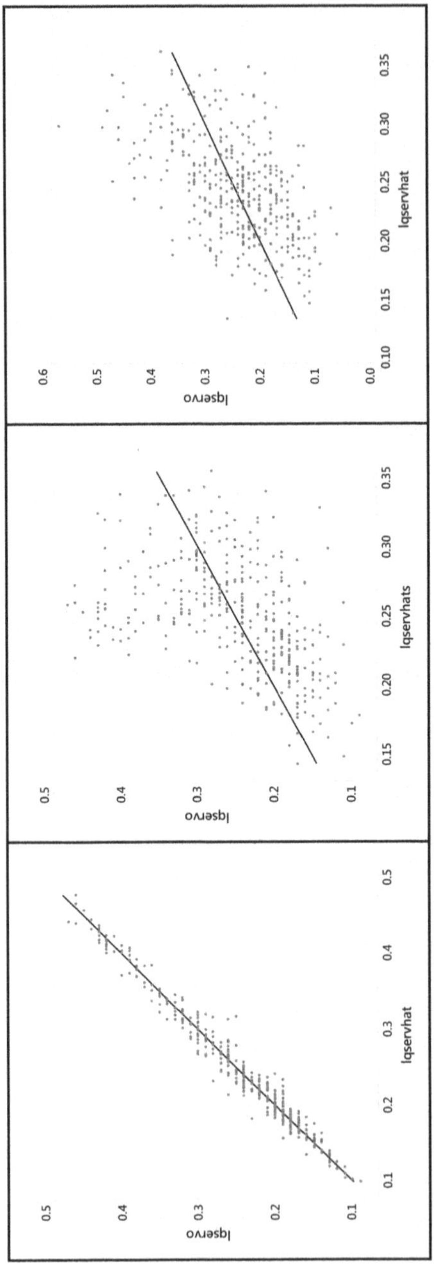

Fig. 4.3 (continued)

Table 4.9 Bivariate regression results for LQs regressed on imputed LQs: space beats time

Statistic	A	I	T	F	S
Temporal prediction					
Intercept	0.0016	−0.0002	−0.0005	−0.0001	−0.0003
Slope	0.9928	1.0010	1.0032	1.0043	1.0018
Pseudo-R^2	0.7863	0.5446	0.3634	0.2968	0.1753

Accordingly, the SSRE term is the only structure in the data remaining that can be exploited for prediction purposes. Scatterplots in the third column of Fig. 4.3 illustrates this space-beats-time situation. Table 4.9 summarizes the corresponding bivariate regression results. Again, the intercept and slope values are close to their expected counterparts.

4.6 Concluding Comments

This chapter presents discussions about the relative importance of temporal and spatial autocorrelation, particularly within the context of forecasting. Space–time data structures are conceptualized in terms of SSRE and SURE components. One contribution to the literature is a sensitivity analysis revealing that relatively good estimates for these two components can be obtained for $t = 1$. Emphasis throughout this chapter is on the tendency for temporal correlation to dominate spatial correlation. Another novel contribution here is a detailed example illustrating spatial correlation dominating temporal correlation, which admittedly is a rare occurrence.

References

Gelfand, A., Kim, H.-J., Sirmans, C., & Banerjee, S. (2003). Spatial modeling with spatially varying coefficient processes. *Journal of the American Statistical Association, 98*, 387–396.

Griffith, D. (2008). A comparison of four model specifications for describing small heterogeneous space-time datasets: Sugar cane production in Puerto Rico, 1958/59–1973/74. *Papers in Regional Science, 87*, 341–356.

Griffith, D. (2013). Estimating missing data values for georeferenced Poisson counts. *Geographical Analysis, 45*, 259–284.

Griffith, D., & Chun, Y. (2014). An eigenvector spatial filtering contribution to short range regional population forecasting. *Economics and Business Letters, 3*, 208–217.

Griffith, D., & Paelinck, J. (2009). Specifying a joint space- and time-lag using a bivariate Poisson distribution. *Journal of Geographical Systems, 11*, 23–36.

Patuelli, R., Griffith, D., Tiefelsdorf, M., & Nijkamp, P. (2011). Spatial filtering and eigenvector stability: Space-time models for German unemployment data. *International Regional Science Review, 34*, 253–280.

Chapter 5
The Spatial Weights Matrix and ESF

5.1 Introduction

The spatial weights matrix (SWM) is a critical component of spatial statistical/econometrics analysis. It articulates the relative locations of areal unit observations and is the source of spatial autocorrelation (SA) and SA analysis, establishing which areal units are spatially correlated. Over the years, considerable discussion has been devoted to misspecification aspects associated with the SWM (see LeSage and Pace 2014). Guidelines have been suggested for constructing it (e.g., Griffith 1996). One useful conclusion from its general assessments is that a slightly misspecified SWM is better than ignoring SA (Griffith and Lagona 1998). A more contemporary concern focuses on whether the specification of a SWM is endogenous or exogenous (Corrado and Fingleton 2012). This chapter addresses techniques for comparing competing SWMs, as well as other SWM features, which reflect upon misspecification issues.

5.2 Spatial Weights Matrix Comparisons

The literature includes a number of different matrix comparison techniques (see Patuelli et al. 2006). Cliff and Ord (1973, p. 166) note that one way to compare SWMs is to construct a likelihood ratio test for each from model specification results. Their approach may have appeal for linear models with normally distributed errors, which is the topic they discuss, but not necessarily for Poisson or binomial

The content of this chapter was compiled and developed for a graduate spatial statistics/econometrics course taught at the University of Texas at Dallas during the Spring semester of 2015.

D. A. Griffith, J. H. P. Paelinck, *Morphisms for Quantitative Spatial Analysis*,
Advanced Studies in Theoretical and Applied Econometrics 51,
https://doi.org/10.1007/978-3-319-72553-6_5

regression models. Meanwhile, likelihood ratio tests also have been promoted for the problem of comparing the equality of two or more covariance matrices (Manly and Rayner 1987), which again often requires an assumption of normality. In these latter cases, ratios are based upon the determinants or the traces of the matrices being compared. This approach is not suited for SWMs, whose traces are zero, and whose determinants may be zero or negative. Furthermore, these measures are single summary quantities for which the same value can result from various combinations of matrix elements. The Frobenius norm (i.e., the square root of the sum of squared matrix elements) is another such summary quantity used to compare two matrices.

A second approach is to calculate the correlation between the respective (row, column) elements of two matrices. A serious weakness of this comparison technique arises from the sparseness of certain SWMs, in which cases each contains a large percentage of paired zero elements, inflating this similarity measure for the two matrices; for a planar surface partitioning, such a SWM would contain $n^2 - 6(n - 2)$ zeroes. Consider n-by-n binary 0/1 SWMs, one defined with rook adjacency and its competitor defined with queen adjacency (Table 5.1). The asymptotic correlation, as $n$goes to infinity, for the square tessellation rook and queen SWMs is $\frac{1}{\sqrt{2}} \approx 0.70711$. The asymptotic correlation for the complete planar tessellation (i.e., the maximum number of planar adjacencies is present) is $\frac{1}{\sqrt{k+1}}$, where $k \in [0, 1]$. Frequently in practice, the number of additional 1s in queen versus rook definitions of adjacency for an irregular surface partitioning is about 10%, which results in an asymptotic correlation of 0.95346. When k reaches 100% (i.e., a doubling of the number of 1s), this correlation is 0.70711. The asymptotic correlation for the K- and K+1-nearest neighbors (NN) adjacency is $\sqrt{K/(K+1)}$. Accordingly, going from five to six NN results in a correlation of 0.91287; going from $n - 2$ to $n - 1$ (the maximum number of NN possible) results in an asymptotic correlation of 1 (Table 5.1). In other words, the correlation coefficient for two binary SWMs tends to be quite large in most cases and, hence, fails to differentiate effectively between two dissimilar SWMs.

For binary 0/1 SWMs, the three preceding summary statistics relate to their eigenvalues: the determinant of a SWM equals the product of its eigenvalues, the trace of a SWM equals the sum of its eigenvalues, and the Frobenius norm equals the square root of the sum of a SWM's squared eigenvalues. In other words, the eigenvalues of two SWMs, say matrices $\mathbf{C}_1$ and $\mathbf{C}_2$, furnish a comparison criterion to differentiate between them. Garcia (2012), again with reference to covariance matrices, proposes using the full sets of n eigenvalues to differentiate between two matrices.[1] The technique here resembles a quantile plot: if a scatterplot of two sets of rank-ordered eigenvalues depicts a straight line, then the two matrices are the same. The question needing to be answered asks how close the scatterplot needs to

[1]Determining which SWMs are uniquely determined by their spectra is a very difficult problem. Only a few SWMs have been shown to have this property, with a few also having been shown to be cospectral. Van Dam and Haemers (2003) suggest that almost all SWMs have this property.

Table 5.1 Comparisons of pairs of 0s and 1s in binary SWMs constructed with different adjacency definitions

		Rook adjacency					K+1-Nearest neighbors (NN)	
		Square tessellation		Complete planar				
Cell entry		**1**	**0**	**1**	**0**		**1**	**0**
1	Queen adjacency	$4(n - \sqrt{n})$	$4(n - 2\sqrt{n} + 1)$	$6(n - 2)$	$k6(n - 2)$	K NN	Kn	n
0		0	$n^2 - 4(2n - 3\sqrt{n} + 1)$	0	$n^2 - (k + 1)6(n - 2)$		0	$n^2 - (K + 1)n$

Note: K denotes the positive integers; k denotes a value in the interval [0, 1]. Bold font denotes the cell entry in an n-by-n binary 0/1 SWM

Table 5.2 Gamma fitting critical values determined by a simulation experiment with 100,000 replications and a sample size of 380

Significance level	r	$\lvert 1 - b \rvert$
10%	0.985720	0.315475
5%	0.981570	0.384665
1%	0.970015	0.548395

be to a straight line; in other words, what is the threshold correlation coefficient for the two sets of eigenvalues?

Eigenvalues were calculated for the rook adjacency SWM. Because when adjusted such that the minimum is subtracted from each eigenvalue, and then this difference is divided by the maximum minus the minimum eigenvalue, which converts the set of n eigenvalues to the interval [0, 1], the adjusted eigenvalues seem like they should be described by a beta distribution. But all diagnostic probabilities for a beta distribution description are less than 0.0001. Past experience suggests that these adjusted eigenvalues are best described by a truncated gamma distribution; diagnostic probabilities for this gamma description are reasonable: P(Kolmogorov–Smirnov) = 0.172; P(Cramer–von Mises) = 0.123; and P(Anderson–Darling) = 0.046.

The fitted gamma distribution has a scale parameter of 0.2 and a shape parameter of 1.6. A simulation experiment with 100,000 replications, each of which has 380 adjusted values drawn at random from a gamma distribution with the preceding parameters, yields the critical value correlation (r) and bivariate regression slope (b) coefficients appearing in Table 5.2.

These critical values can be used for scatterplot comparisons of SWMs for the Polish powiat surface partitioning. One implication of these results is that a queen adjacency SWM with more than 2.9–6.3% additional ones (for $n = 380$; this percentage also is a function of n) is significantly different from a rook adjacency SWM. Thus, because most queen adjacency SWMs for irregular square tessellations have approximately 10% more ones, they actually are statistically different SWMs.

5.2.1 Some Binary SWM Comparisons

The rook and queen adjacency definitions for the Polish powiaty yield exactly the same binary SWM (i.e., no pair of powiaty share only a single boundary point). Consequently, the correlation between the two sets of eigenvalues is 1 (Fig. 5.1a), which is greater than 0.985720: the sets of eigenvalues furnish no evidence to indicate that the two SWMs are not essentially the same. The average number of neighbors with these two adjacency definitions is 5.3, suggesting comparisons with 4- and 6-nearest neighbor (NN) adjacency SWMs; the rook SWM also is compared with the 8 NNs SWM.

The respective correlation coefficients for the 4-, 6-, and 8-NNs SWMs are 0.993986, 0.99363, and 0.98217 (Fig. 5.1b–d), indicating that only the 8-NNs

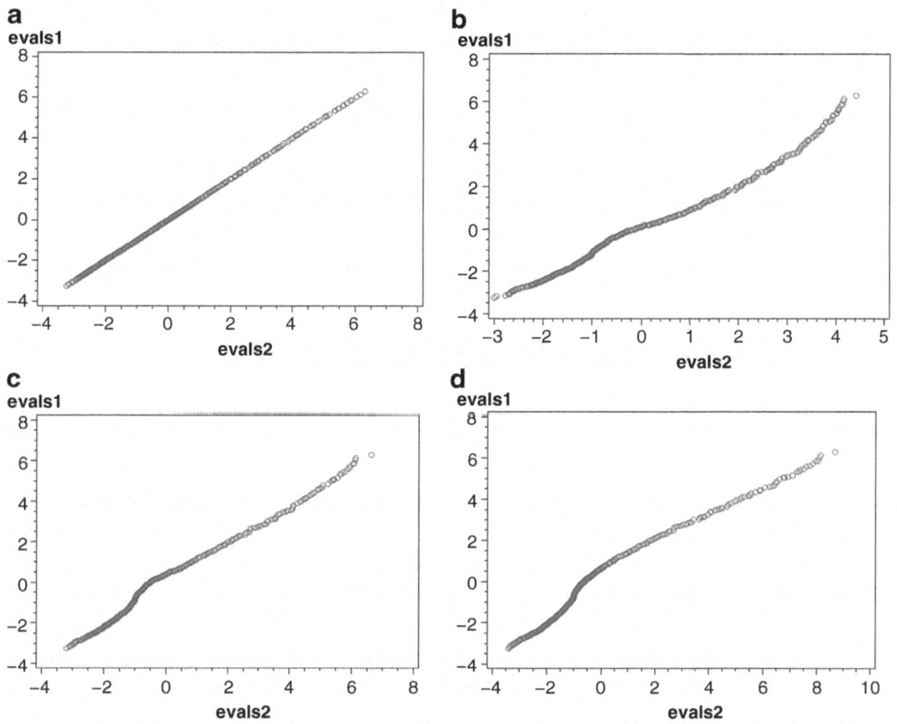

Fig. 5.1 Scatterplots of rank-ordered eigenvalues for pairs of SWMs. Top left (**a**): rook and queen adjacency SWMs. Top right (**b**): rook and 4 NNs adjacency SWMs. Bottom left (**c**): rook and 6 NNs adjacency SWMs. Bottom right (**d**): rook and 8 NNs adjacency SWMs

SWM differs somewhat from the rook adjacency SWM. The Frobenius norms fail to support this result: 44.8 for the rook SWM, 37.2 for the 4-NNs SWM, 45.8 for the 6-NNs SWM, and 53.0 for the 8-NNs SWM. Because the correlation coefficient furnishes a good measure of the linearity of the scatterplot of eigenvalues, and hence furnishes valuable information about the similarity of two SWMs, it can be combined with the Frobenius norms to give a better diagnostic; the ratio of the Frobenius norms times the corresponding correlation coefficient is the slope of the bivariate regression line for the corresponding scatterplot. Accordingly, the respective bivariate regression slope coefficients for the 4-, 6-, and 8-NNs are 1.19708, 0.97160, and 0.83030. Delaunay triangulation furnishes another SWM that is similar to the rook adjacency SWM: $r = 0.99725$, and $|1 - b| = |1 - 0.94289| = 0.05711$. These diagnostics reveal that each of the three NN SWMs, as well as the Delaunay triangulation SWM, are similar to the rook adjacency SWM (this corroborates arguments by Griffith and Lagona 1998).

The 1-NN SWM furnishes a contrast to these five preceding cases; it is significantly different from the rook adjacency SWM: $r = 0.93114$, and $|1 - b| = |1 - 2.39980| = 1.33980$. Its scatterplot is concave upward. A 1-NN SWM tends to have too few adjacencies, given that the rook adjacency case has an average of 5.3

neighbors. Rather than too few, a SWM specification could have too many adjacencies. Consider the 15-NNs SWM (its average is 15 adjacencies, which is substantially greater than 5.3); it is significantly different from the rook adjacency SWM: $r = 0.91937$, and $|1 - b| = |1 - 0.56544| = 0.43456$ (which is significant at the 5%, but not the 1%, level). Its scatterplot is concave downward. These findings suggest that more research needs to address the issue of comparing SWMs. For example, the principal eigenvalues indicate that 4- and 8-NN SWMs are not very similar to the rook adjacency SWM ($\lambda_1 = 4.4$ and 8.7 vs. 6.3).

5.2.2 Some Row-Standardized SWM Comparisons

Spatial scientists often employ other SWM definitions, some of which involved row standardization (i.e., sum the elements in a row, and then divide each entry in that row by its row sum); all SMW row sums become 1. One appealing feature of row-standardized SWMs is that the value of an attribute variable at a given location is a function of the average of its neighboring attribute values. In addition, row standardization allows a SWM specification to be based upon distances between locations (e.g., distances between areal unit centroids) rather than areal unit boundary contiguity.

The preceding binary 0/1 rook adjacency SWM was converted to its row-standardized counterpart, matrix **W**. This modification immediately rescales its principal eigenvalue to 1. Figure 5.2a portrays the scatterplot comparison of this SWM with that based upon the following distance formula:

$$w_{ij} = \frac{\exp(-\lambda d_{ij})}{\sum_{j=1}^{n} \exp(-\lambda d_{ij})}, \tag{5.1}$$

where d_{ij} denotes the distance between locations i and j, exp is the antinatural logarithm, and λ is the distance decay parameter. With the parameter $\lambda = 0.65$, this specification is not statistically different from the row-standardized binary 0/1 specification: $r = 0.98337$, and $|1 - b| = |1 - 1.20332| = 0.20332$. As the distance decay parameter λ increases, the range of the spatial field shrinks. Meanwhile, the following is a more popular distance formula:

$$w_{ij} = \frac{1/d_{ij}^{\lambda}}{\sum_{j=1}^{n} 1/d_{ij}^{\lambda}}. \tag{5.2}$$

Common values for λ, in practice, are 1 (i.e., inverse distance) and 2 (i.e., inverse distance squared). Figure 5.2b portrays the eigenvalue scatterplot for definition (5.2) with $\lambda = 1$; it does not resemble a straight line. This specification is statistically different from the row-standardized binary 0/1 specification: $r = 0.74312$, and $|1 - b| = |1 - 3.53428| = 2.53428$. Its scatterplot more closely resembles a logistic curve. Increasing the distance decay exponent to 2 (Fig. 5.2c) yields a specification

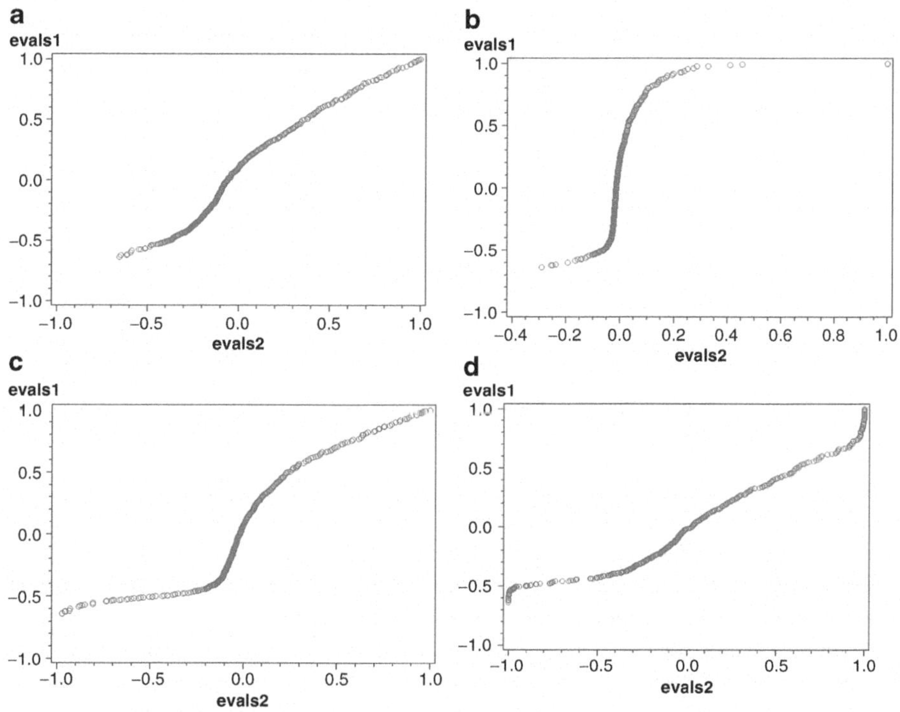

Fig. 5.2 Matrix **W** eigenvalue comparisons. Top left (**a**): rook and negative exponential distance SWMs. Top right (**b**): rook and inverse distance SWMs. Bottom left (**c**): rook and inverse distance squared SWMs. Bottom right (**d**): rook and a 3.5 exponent inverse distance SWMs

that may or may not be statistically different from the row-standardized binary 0/1 specification: $r = 0.91396$, and $|1 - b| = |1 - 1.15073| = 0.15073$. Its scatterplot is somewhat, but not approximately, linear. Finally, increasing the distance decay exponent to 3.5 (Fig. 5.2c) renders a specification that is not statistically different from the row-standardized binary 0/1 specification: $r = 0.97427$, and $|1 - b| = |1 - 0.76676| = 0.23324$.

5.2.3 Variance Stabilizing Standardization

Tiefelsdorf et al. (1999) present an alternative to the preceding two specifications. They highlight that the denominator for the row-standardized binary 0/1 SWM can have the following general form:

$$w_{ij} = \frac{c_{ij}}{\left(\sum_{j=1}^{n} c_{ij}\right)^{\gamma}}, \tag{5.3}$$

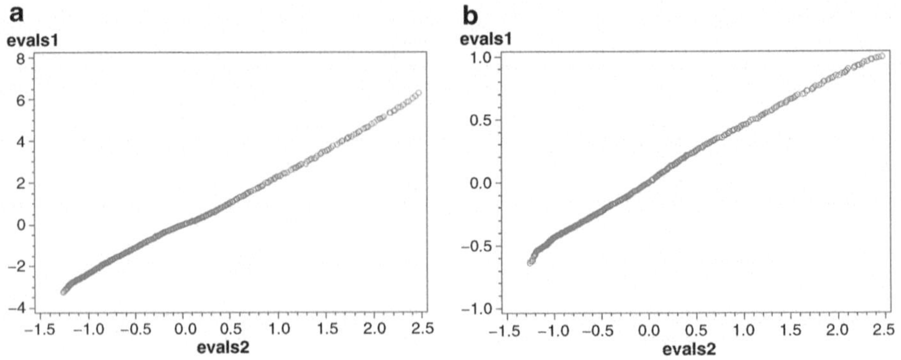

Fig. 5.3 Matrix **S** comparisons. Left (**a**): rook **C** and **S** SWMs. Right (**b**): rook **W** and **S** SWMs

where $0 \leq \gamma \leq 1$. Matrix **C** results from $\gamma = 0$; matrix **W** results from $\gamma = 1$. The matrix **C** specification emphasizes interior areal units at the expense of border areal units. The matrix **W** specification emphasizes border areal units at the expense of interior areal unit. This trade-off roughly balances with $\gamma = 0.5$: emphasis on interior and border areal units is approximately the same.

Figure 5.3 portrays the eigenvalue scatterplots comparing matrices **C** and **S** (Fig. 5.3a) and matrices **W** and **S** (Fig. 5.3b). Both plots are very linear, respectively, producing correlations of 0.99890 and 0.99883; the implication is that the pairs of matrices essentially are equivalent. However, the slopes for both plots suggest a significant difference: for matrices **C** and **S**, $|1 - b| = |1 - 2.36298544| = 1.36298$, and, for matrices **W** and **S**, $|1 - b| = |1 - 0.43924| = 0.56076$. In other words, the diagnostics yield inconclusive results. Part of this discrepancy is attributable to, and indexed by, the principal eigenvalues of the three matrices: 6.3 for matrix **C**, 2.5 for matrix **S**, and 1.0 for matrix **W**.

A primary conclusion here is that more research needs to be completed about matrix comparisons employing the full sets of eigenvalues.

5.3 Comparisons of Spatial Weights Matrix Eigenvectors

Eigenvector spatial filtering (see Sect. 5.2.2) employs eigenvectors extracted from a SWM as synthetic variates to account for SA in a model specification. Specifically, the eigenvectors are extracted from the matrix

$$\left(\mathbf{I} - \mathbf{1}\mathbf{1}^{\mathrm{T}}/n\right)\mathbf{C}\left(\mathbf{I} - \mathbf{1}\mathbf{1}^{\mathrm{T}}/n\right). \tag{5.4}$$

The geographic distributions of these eigenvectors portray distinct map patterns that visualize particular natures and degrees of SA, which are quantified globally by their associated eigenvalues. One important question asks whether or not different SWM specifications result in different ESFs. In statistical parlance, this is a

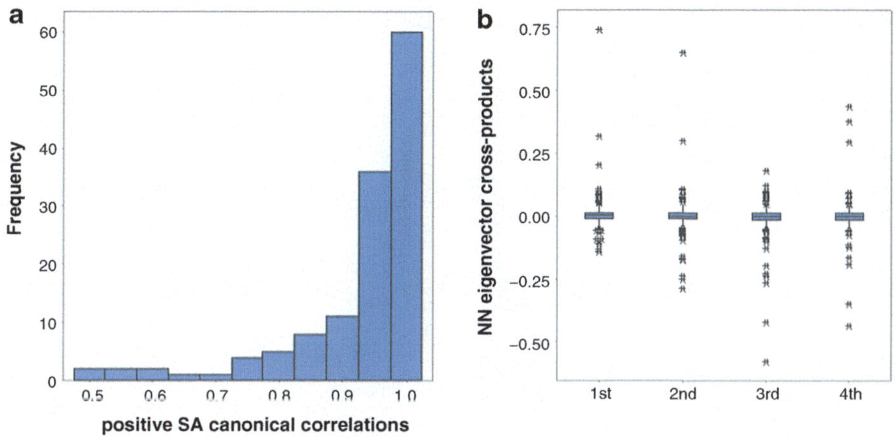

Fig. 5.4 CCA results for positive SA eigenvectors. Left (**a**): histogram of 132 canonical correlations. Right (**b**): boxplots of 6-NN SWM eigenvector cross-product values for the first four rook adjacency SWM eigenvectors

canonical correlation analysis[2] (CCA) problem: do latent dimension span the eigenvectors of two SWMs for the same surface partitioning? Because each set of eigenvectors is mutually orthogonal and uncorrelated, this question asks whether or not subsets of eigenvectors in each of the two sets are equivalent.

The rook adjacency and 6 NNs SWMs, respectively, have 161 and 132 eigenvectors representing negligible to marked positive SA. A CCA of these two sets of eigenvectors produces 132 linear combination pairs with most canonical correlations being very close to 1 (Fig. 5.4a). The linear combinations do not necessarily indicate a close correspondence between individual eigenvectors; eigenvector cross-products reveal some dominant linkages combined with a mixture of weak linkages (e.g., Fig. 5.4b). In other words, the two sets of eigenvectors partition the SA mathematical space in a different, but cumulatively the same, way.

The row-standardized rook adjacency [Eq. (5.3), $\gamma = 1$] and row-standardized negative exponential distance (Eq. 5.1) SWMs, respectively, have 162 and 138 eigenvectors representing negligible to marked positive SA. A CCA of these two sets of eigenvectors produces 138 linear combination pairs with most canonical correlations being very close to 1 (Fig. 5.5a). Again, the linear combinations do not necessarily indicate a close correspondence between individual eigenvectors; eigenvector cross-products reveal some dominant linkages combined with a mixture of weak linkages (e.g., Fig. 5.5b). In other words, as before, the two sets of

[2]Canonical correlation analysis is a multivariate statistical technique for constructing linear combinations, one from each of two sets of attribute variables measured with the same areal units, such that each pair of linear combinations is maximally correlated. In doing so, it identifies dimensions that span two sets of variables.

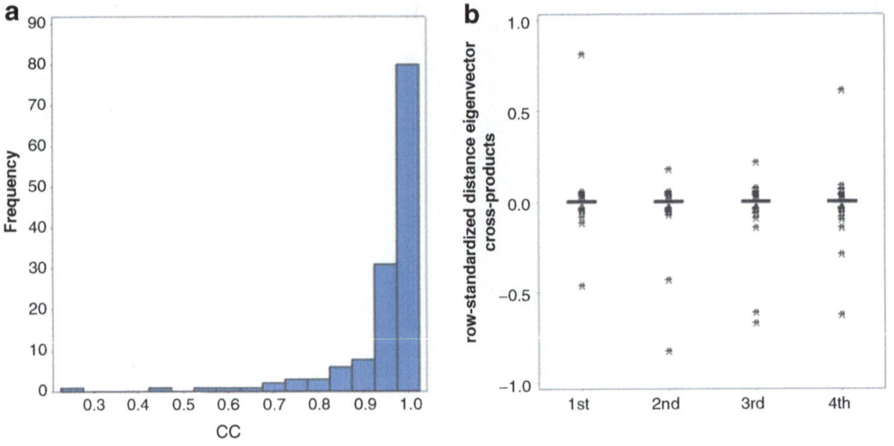

Fig. 5.5 CCA results for positive SA eigenvectors. Left (**a**): histogram of 138 canonical correlations. Right (**b**): boxplots of row-standardized negative exponential distance SWM eigenvector cross-product values for the first four row-standardized rook adjacency SWM eigenvectors

eigenvectors partition the SA mathematical space in a different, but cumulatively the same, way.

In conclusion, evidence presented in this section supports the notion that eigenvectors for different SWM specifications for the same set of areal units essentially are equivalent in total. The expectation is that some SWM eigenvectors partition the SA mathematical space more efficiently and parsimoniously than others.

5.4 Competing Model Specifications: Spatial Autoregressions and ESFs

Eigenvector spatial filtering is based upon spatial autoregression. The following algebraic analysis highlights relationships between these two types of model specifications:

$$\begin{aligned}
\mathbf{Y} &= \rho\mathbf{WY} + \mu(\mathbf{I} - \rho\mathbf{W})\mathbf{1} + \boldsymbol{\varepsilon} \\
&= \rho\mathbf{WY} + \mu(1 - \rho)\mathbf{1} + \boldsymbol{\varepsilon} \\
\mathbf{Y} - \rho\mathbf{WY} &= \mu(1 - \rho)\mathbf{1} + \boldsymbol{\varepsilon} \\
(\mathbf{I} - \rho\mathbf{W})\mathbf{Y} &= \mu(1 - \rho)\mathbf{1} + \boldsymbol{\varepsilon} \\
\mathbf{Y} &= \mu\mathbf{1} + (\mathbf{I} - \rho\mathbf{W})^{-1}\boldsymbol{\varepsilon} \\
&= \mu\mathbf{1} + \left[\sum_{k=0}^{n}(\rho\mathbf{W})^k\right]\boldsymbol{\varepsilon} = \mu\mathbf{1} + (\boldsymbol{\varepsilon} + \boldsymbol{\xi})
\end{aligned}$$

Eigenvector spatial filtering captures the SA in the $\boldsymbol{\xi}$ term. It essentially decomposes the SWM into its eigenfunctions and retains only those eigenvectors that exhibit a significant relationship with a response variable, Y.

If a response variable exhibits positive SA, then attention can be restricted to those eigenvectors associated with positive SA. Consider the Polish $LQ_{agriculture}$ for 2013 and the rook adjacency SWM; the expression (5.4) modified version of this SWM has 161 positive, 1 zero, and 218 negative SA eigenvectors. The SA parameter estimate for a spatial simultaneous autoregressive (SAR) model specification[3] is $\widehat{\rho} = 0.82$, with a pseudo-R^2 of 0.63. Its ESF counterpart, using a significance level selection criterion of 0.01, contains 23 eigenvectors and has a R^2 of 0.66. The absolute value of the linear combination $\mathbf{E}_{218}^{\mathrm{T}}\mathbf{Y}$, where T denotes the matrix transpose operator and $\mathbf{E}_{218}$ is the matrix of 218 eigenvectors related to negative SA, ranges from 0.0 to 33.8. The eigenvector with an affiliated eigenvalue of zero is a vector proportional to **1** and hence relates to the model's intercept term. The absolute value of the linear combination $\mathbf{E}_{138}^{\mathrm{T}}\mathbf{Y}$ of the 138 positive SA eigenvectors not selected to construct the ESF ranges from 0.1 to 39.0. Finally, the absolute value of the linear combination $\mathbf{E}_{23}^{\mathrm{T}}\mathbf{Y}$ of the 23 positive SA eigenvectors selected to construct the ESF ranges from 26.7 to 382.5; here, five eigenvector cross-calculations overlap with the other two ranges, but four of them have moderate Moran Coefficients. The strong tendency here is for relatively large $\mathbf{E}^{\mathrm{T}}\mathbf{Y}$ values to identify prominent eigenvectors and relatively $\mathbf{E}^{\mathrm{T}}\mathbf{Y}$ small values to identify eigenvectors that can be ignored.

Griffith (2012) contrasts this ESF situation with the individual SAR terms. For the Polish example analyzed here, the negative SA component of the $\rho\mathbf{WY}$ counterpart to the ESF is similar: the individual areal unit terms range from 0.0 to 33.1. The positive SA decomposition is not so clear. Using the threshold of 26.7 established in the preceding paragraph by the ESF analysis, both specifications would discard 123 of the eigenvectors, both emphasize 11 of the eigenvectors, the ESF emphasizes 12 eigenvectors ignored by the SAR model, and the SAR model emphasizes 14 eigenvectors ignored by the ESF. The overall impact of these differences is that the SAR model has to accommodate more noise. One outcome is that if the level of significance for eigenvector selection increases to 0.05, the ESF specification R^2 increases to 0.75 (the ESF contains 45 eigenvectors); if the level of significance increases to 0.10, the R^2 increases to 0.79 (the ESF contains 61 eigenvectors).

5.5 Concluding Comments

SWMs merit considerably more study in spatial statistics/econometrics. Spatial scientists need to have robust ways to compare them. The eigenvalue scatterplot presented in this chapter shows considerable promise as such a comparative tool. Regardless, if a SWM specification is reasonable, modest deviations from it appear not to be problematic. But sensitivity analyses and tolerance intervals need to be

[3]Also known as a spatial error model specification in the spatial econometrics literature.

completed in order to create proper SWM specification guidelines. One contribution of this chapter is that especially the inverse distance SWM seems to be a poor specification. The appropriate distance exponent appears to be much greater than 1, and perhaps even greater than 2. Furthermore, a negative exponential distance specification may well be more suitable. Finally, one useful advantage ESFs appear to have over spatial autoregressive models is that they set aside considerable noise from a spatial analysis by employing only the data-relevant subset of eigenvectors that can be extracted from a SWM.

References

Cliff, A., & Ord, J. (1973). *Spatial autocorrelation.* London: Pion.

Corrado, L., & Fingleton, B. (2012). Where is the economics in spatial econometrics? *Journal of Regional Science, 52*, 210–239.

Garcia, C. (2012). A simple procedure for the comparison of covariance matrices. *BMC Evolutionary Biology, 12*, 222.

Griffith, D. (1996). Some guidelines for specifying the geographic weights matrix contained in spatial statistical models. In S. Arlinghaus & D. Griffith (Eds.), *Practical handbook of spatial statistics* (pp. 65–82). Boca Raton, FL: CRC Press.

Griffith, D. (2012). Spatial statistics: A quantitative geographer's perspective. *Spatial Statistics, 1*, 3–15.

Griffith, D., & Lagona, F. (1998). On the quality of likelihood-based estimators in spatial autoregressive models when the data dependence structure is misspecified. *Journal of Statistical Planning and Inference, 69*, 153–174.

LeSage, J., & Pace, R. (2014). The biggest myth in spatial econometrics. *Econometrics, 2*, 217–249.

Manly, B., & Rayner, J. (1987). The comparison of sample covariance matrices using likelihood ratio tests. *Biometrika, 74*, 841–847.

Patuelli, R., Griffith, D., Tiefelsdorf, M., & Nijkamp, P. (2006). *The use of spatial filtering techniques: The spatial and space-time structure of German unemployment data.* Discussion Paper No. 06-049/3. Amsterdam: Tinbergen Institute.

Tiefelsdorf, M., Griffith, D., & Boots, B. (1999). A variance stabilizing coding scheme for spatial link matrices. *Environment and Planning A, 31*, 165–180.

van Dam, E., & Haemers, W. (2003). Which graphs are determined by their spectrum? *Linear Algebra & Its Applications, 373*, 139–162.

Chapter 6
Clustering: Spatial Autocorrelation and Location Quotients

6.1 Introduction

Clustering in spatial statistics refers to at least two manifestations: within areal unit concentration of phenomena and global spatial autocorrelation trends that include local hot spots of relatively high or relatively low values. Content in this chapter illustrates these notions in terms of location quotients (LQs) by describing the geographic distribution of annual employment by five general categories across Poland for the powiat (NUTS4 ≈ county; $n = 380$; see Fig. 1.1a) spatial resolution during 2006–2013. Covariates in this description are working age population density (Fig. 1.1c), and percentage of working age population (Fig. 1.1d). This chapter extends the existing literature by positing a LQ specification that not only produces confidence intervals for measures, but also accounts for spatial autocorrelation in terms of prevailing map pattern.

The isomorphism of interest here is the cluster: a LQ cluster is within an areal unit, whereas a spatial LQ cluster is a concentration of similar neighboring LQ values. This chapter relates to the spatial econometrics outlined in Chap. 17, which alludes to spatial LQ clusters. Although the preferred geographic situation is for concentrations of high LQ values, another type of cluster is the concentration of low LQ values. Emphasis also on this latter type of cluster is a contribution of spatial statistics.

6.2 Location Quotients

A LQ is an index quantifying how concentrated a particular employment category is in an areal unit compared to its parent nation (or other coarser geographic resolution context). Its measure is given by

D. A. Griffith, J. H. P. Paelinck, *Morphisms for Quantitative Spatial Analysis*, Advanced Studies in Theoretical and Applied Econometrics 51,
https://doi.org/10.1007/978-3-319-72553-6_6

$$(E_{C,i}/E_{T,i})/(E_{C,N}/E_{T,N}),$$

where $E_{C,i}$ denotes the total employment in category C in areal unit i, $E_{T,i}$ denotes the total employment in areal unit i, $E_{C,N}$ denotes the total employment in category C in nation N (or some other reference geographic landscape), and $E_{T,N}$ denotes the total employment in nation N (or some other reference geographic landscape). A LQ may be interpreted as follows:

LQ > 1: its areal unit has a higher than the national average concentration of a given employment category (specialization; employment clustering).
LQ < 1: its areal unit has a lower than the national average concentration of a given employment category.

A LQ actually is a percentage of local employment by category (its numerator) standardized by the national percentage of employment in its given category (its denominator). As such, it can be treated as a rescaled binomial random variable (RV). This approach taken by Moineddin et al. (2003) and Beyene and Moineddin (2005) allows confidence intervals to be attached to LQs by casting them as binomial RVs within a generalized linear model (GLM) framework, but these authors fail to specify an auto-binomial or other spatial version of their model. Accordingly, their analysis fails to account for spatial autocorrelation in geographic distributions of LQs.

6.3 The Multivariate Space–Time Structure of Polish LQs: 2006–2013

Subjecting the 360 spatial time series to a principal components analysis yields two data dimensions for the space–time data cube: one accounting for 56.4% and the other accounting for 31.4%, of the varian (Fig. 6.1). Employment category A LQs are negatively related to both of these components, with roughly the same loadings. In addition, employment categories T, S, and O LQs correlate positively and strongly to the first of these components, whereas employment category I LQs correlate positively and strongly to the second of these components. Both principal components contain weak-to-moderate, positive spatial autocorrelation [respective Moran Coefficients (MC) of 0.26 and 0.45].

Table 6.1 summarizes the individual annual multivariate LQ structures. Of note is that: (1) the multivariate structure is stable over the given time horizon, (2) rural activity employment concentration negatively relates to other employment concentrations and equally spans both factors, and (3) non-industrial activity employment concentrates in a way that is independent of industrial activity employment concentration.

Table 6.2 reveals that (1) these employment data display weak-to-moderate positive spatial autocorrelation (2) because all employment categories but one have an MC + GR that is not approximately 1, these data appear to be ill-behaved (e.g.,

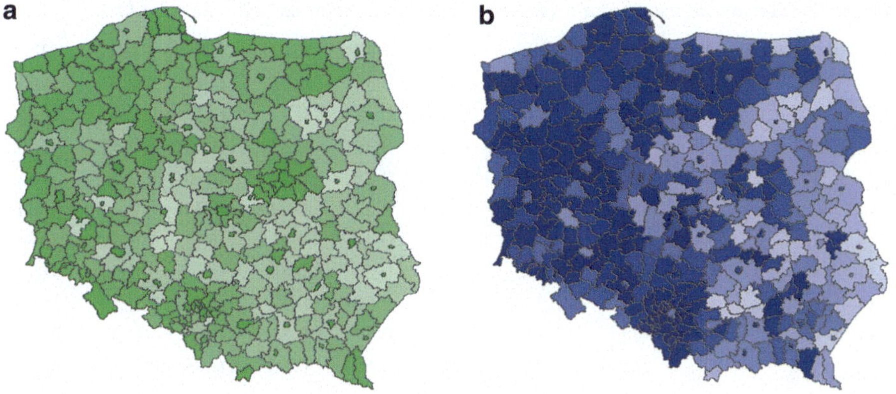

Fig. 6.1 Intensity of color is directly proportional to the LQ values. *Left* (**a**): Principal component I. *Right* (**b**): Principal component II

Table 6.1 LQ factor analytic results for individual years

	Year							
Employment category	2006	2007	2008	2009	2010	2011	2012	2013
FACTOR I (varimax rotated principal component I)								
LQ-A	−0.71	−0.70	−0.71	−0.73	−0.72	−0.73	−0.73	−0.73
LQ-I	0.08	0.06	0.06	0.06	0.05	0.04	0.04	0.04
LQ-T	0.85	0.85	0.83	0.84	0.84	0.84	0.84	0.83
LQ-S	0.88	0.88	0.88	0.88	0.88	0.88	0.87	0.88
LQ-O	0.90	0.89	0.90	0.91	0.91	0.91	0.90	0.90
% variance accounted	56.6	55.7	55.6	56.7	56.5	56.8	56.2	56.3
FACTOR II (varimax rotated principal component II)								
LQ-A	−0.70	−0.71	−0.70	−0.68	−0.69	−0.68	−0.68	−0.68
LQ-I	0.99	0.99	0.99	0.99	0.99	0.99	0.99	0.99
LQ-T	0.27	0.28	0.28	0.28	0.29	0.27	0.26	0.26
LQ-S	0.12	0.10	0.11	0.10	0.07	0.05	0.05	0.04
LQ-O	0.09	0.09	0.07	0.06	0.07	0.07	0.07	0.08
% variance accounted	31.5	31.7	31.5	30.9	31.0	30.7	30.5	30.4

non-normal); (3) the logarithm is the best Box-Cox power transformation for most, although it fails to align many of them closely with a normal distribution (normal quantile plots confirm this conjecture); and (4) LQs can be described as spatial binomial RVs.

Table 6.2 The annual nature and degree of spatial autocorrelation (MC) for the LQs

Employment category	2006	2007	2008	2009	2010	2011	2012	2013
LQ-A MC	0.53	0.53	0.53	0.54	0.53	0.53	0.53	0.53
LQ-I MC	0.47	0.46	0.47	0.48	0.47	0.48	0.47	0.45
LQ-T MC	0.32	0.32	0.34	0.36	0.35	0.36	0.37	0.38
LQ-S MC	0.25	0.25	0.24	0.25	0.24	0.26	0.26	0.25
LQ-O MC	0.27	0.25	0.25	0.27	0.24	0.24	0.24	0.24
LQ-A GR	0.42	0.43	0.42	0.41	0.41	0.41	0.41	0.40
LQ-I GR	0.56	0.58	0.56	0.56	0.56	0.55	0.55	0.57
LQ-T GR	0.52	0.51	0.51	0.50	0.52	0.52	0.53	0.52
LQ-S GR	0.56	0.56	0.56	0.57	0.56	0.58	0.57	0.56
LQ-O GR	0.57	0.57	0.55	0.54	0.56	0.55	0.54	0.54

Notes: GR denotes the Geary Ratio

6.4 Spatial Autocorrelation and LQs

A small literature exists about LQs and spatial autocorrelation, although it does not address explicitly accounting for spatial autocorrelation in the calculation of LQs. Carroll et al. (2008) compare LQs with the Getis-Ord G_i* statistic. Feng and Ji (2011) describe LQs in terms of local indices of spatial autocorrelation (LISAs), which also is done in this chapter. Cromley and Hanink (2012) incorporate spatial autocorrelation into LQs via the Getis-Ord G_i* statistic, which is more in keeping with this chapter's perspective.

Following Moineddin et al. (2003) and Beyene and Moineddin (2005),

$$\mathrm{LQ} = \left({E_{\mathrm{C},i}}/{E_{\mathrm{T},i}}\right) / \left({E_{\mathrm{C,N}}}/{E_{\mathrm{T,N}}}\right)$$

$$\left({E_{\mathrm{C,N}}}/{E_{\mathrm{T,N}}}\right) \equiv \text{constant}$$

$$\therefore \left({E_{\mathrm{C},i}}/{E_{\mathrm{T},i}}\right) \Rightarrow E_{\mathrm{C},i} \sim \text{binomial}(E_{\mathrm{T},i}, p_i)$$

where $p_i = \frac{1}{1+\exp(\alpha + X_i\beta + \xi_i)}$, and ξ_i represents spatial autocorrelation (i.e., a spatially structured random effects term; SSRE) plus a spatially unstructured random effects (SURE) term, which may well relate to omitted variables.[1] Here spatial autocorrelation will be included with an eigenvector spatial filter (ESF) representing the SSRE (Griffith 2004).

[1]A random effects term treats a regression residual as a composite that is the sum of two terms, a random observation effect (differences among individual observational units) plus an independent and identically distributed space–time-varying residual error (which links to change over space and time). These two terms cannot be separated without additional information, such as priors for a Bayesian analysis, and repeated measures for a frequentist analysis.

Agglomeration (i.e., localization and urbanization) economies are one source of spatial autocorrelation in spatial economics; these suggest covariates to include in LQ analysis. One possibility is working age population density, which indexes city size and relates to: urbanization economies for employment categories T, S, and O and localization economies for employment category I. Another is percentage of working age population, which indexes labor pool size and tends to relate to: urbanization economies in knowledge intensive economic activities and localization economies in specialized (e.g., industrial) economic environments.

6.5 Spatially Adjusted LQs for Polish Employment

Estimation of REs requires additional information. This additional information can be obtained from prior distributions in a Bayesian analysis or from repeated measures (e.g., a time series for each observation) in a frequentist analysis. The latter approach is employed here.

Table 6.3 summarizes results of the space–time analysis. Of the 379 (one is the intercept term) possible eigenvectors, the SSRE includes 19–31 of them, accounting for 25–50% of the variability in the RE term. The extra-binomial variance is excessive for all employment category LQs, whereas the space–time model

Table 6.3 Estimation results from an ESF specification for an auto-binomial model

	Employment category				
Variable	LQ-A	LQ-I	LQ-T	LQ-S	LQ-O
Working age population density	Rank 1	Rank 3	Rank 1	Rank 1	Rank 3
% working age population	Rank 4	Rank 4	Rank 4	Rank 4	Rank 4
SSRE	Rank 2	Rank 1	Rank 3	Rank 2	Rank 2
SURE	Rank 3	Rank 2	Rank 2	Rank 3	Rank 1
# eigenvectors	31	21	29	25	19
SSRE R^2	0.50	0.39	0.42	0.37	0.27
Deviance statistic	81.1	44.5	32.4	14.4	23.4
Pseudo-R^2	0.98	0.97	0.96	0.95	0.97
Bivariate regression a	0.001	0.010	0.005	0.003	0.003
Bivariate regression b	0.996	0.991	0.992	0.997	0.997
Prob(SW) for RE	0.004	<0.001	0.125	0.023	<0.001
Marginal change in pseudo-R^2s					
Working age population density	0.46	0.32	0.35	0.39	0.04
% working age population	<0.01	<0.01	<0.01	<0.01	0.01
SSRE	0.33	0.38	0.41	0.11	0.22
SURE	0.19	0.27	0.21	0.45	0.70

specification does a very good job of describing the geographic distribution of these LQs over the time horizon (Fig. 6.3). Working age population density tends to be an important covariate, whereas percentage of working age population is a significant but relatively unimportant covariate (the linear correlation between these two covariates is 0.33). Furthermore, both the SSRE (i.e., spatial autocorrelation, suggesting such effects as agglomeration economies) and the SURE (suggesting omitted variables) are important descriptors.

6.6 Space–Time Description of the Polish LQs

As noted in the preceding section, the ESF binomial specification furnishes a very good description of these LQ data (Fig. 6.2).

Table 6.4 summarizes the RE structure. The SSRE collapse onto a single dimension accounting for roughly a third of the variance, again with the rural employment category being negatively related to all others. The SURE has two dimensions reflecting those uncovered in Table 6.1, except that employment category A loads onto only one dimension here. Figure 6.3 portrays the geographic distributions of these SSREs: their MCs range from 0.90 to 0.96 (the maximum MC is 1.175), indicating very strong positive spatial autocorrelation. Figure 6.4 portrays the geographic distributions of the SUREs, whose MCs range from −0.05 to 0.05, indicating the presence of only trace spatial autocorrelation.

Implications here include: (1) SSREs have a single dimension that spans all five employment categories and basically is a map pattern geographic trend from the northwest to the southeast; (2) SUREs have two dimensions, one representing employment categories A and I and the other representing employment categories T, S, and O, and (3) the SURE suggest that an omitted covariate might be an indicator variable differentiating between city and non-city powiaty.

6.7 LQ Spatial Clusters in the Clustering of Employment

Although LQs measure local concentration of an employment type within areal units, another clustering question asks whether or not areal units with similar LQs cluster in a given geographic landscape. LISAs and Getis-Ord statistics allow this question to be addressed. Figure 6.5 portrays the geographic clustering of Polish

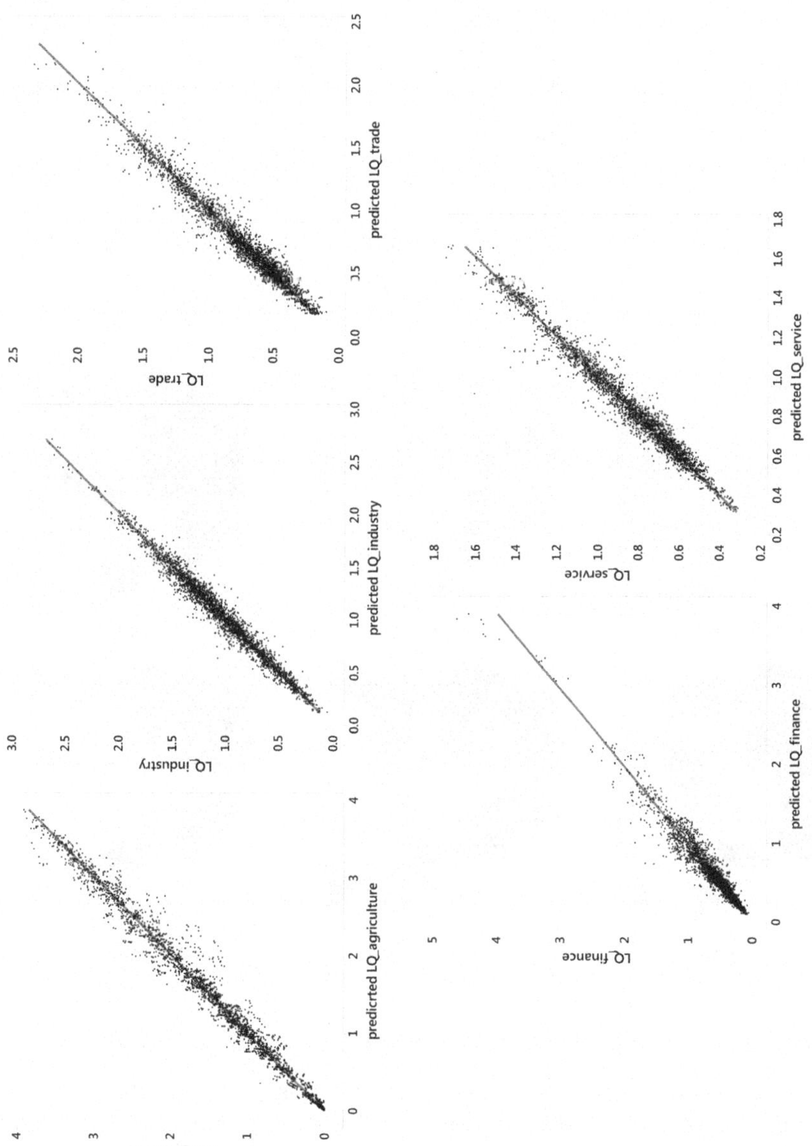

Fig. 6.2 The ESF-binomial space–time scatterplots of observed versus predicted LQs

Table 6.4 The multivariate structure of the REs

Random effect	Factor I	Factor II	Factor III
LQ-A: SSRE	**−0.88**	0.00	0.01
LQ-I: SSRE	**0.75**	−0.05	0.01
LQ-T: SSRE	**0.84**	0.07	−0.06
LQ-S: SSRE	**0.87**	0.05	0.00
LQ-O: SSRE	**0.81**	0.03	0.01
LQ-A: SURE	−0.04	−0.17	**0.91**
LQ-I: SURE	−0.02	−0.36	**−0.83**
LQ-T: SURE	0.00	**0.86**	−0.13
LQ-S: SURE	0.02	**0.90**	0.08
LQ-O: SURE	0.04	**0.92**	0.18
% variance accounted	34.4	25.7	15.8

Notes: Prominent loadings in bold

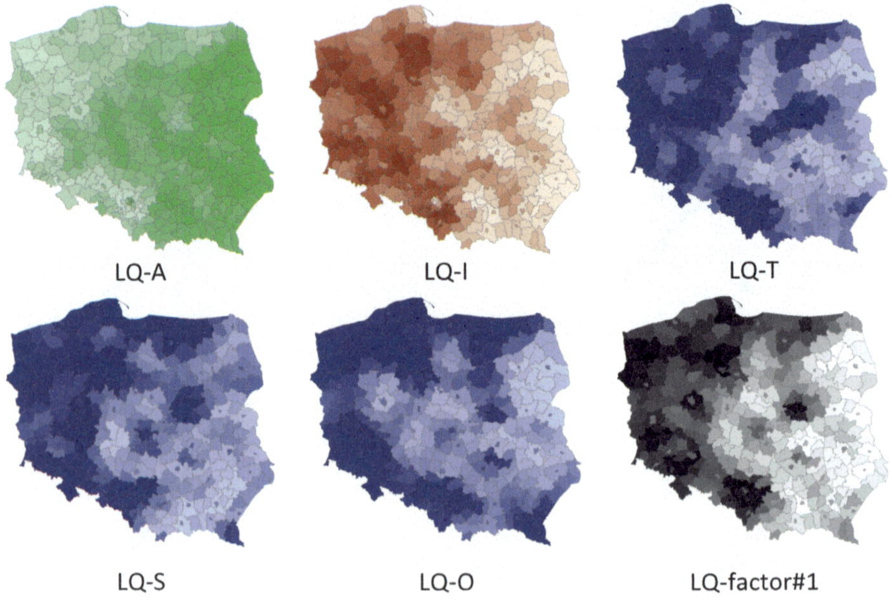

Fig. 6.3 SSREs for the employment categories

LQs. Principal urban regions constitute the hot spots, whereas the more rural eastern part of the county constitutes a massive cold spot (this coincides with the global LQ map patterns; Fig. 6.1). Warsaw is a single anolomy in the national trend. The accompanying Moran scatterplot (Fig. 6.6) reveals only HH and LL geographic clusters; no HL or LH geographic clusters are present.

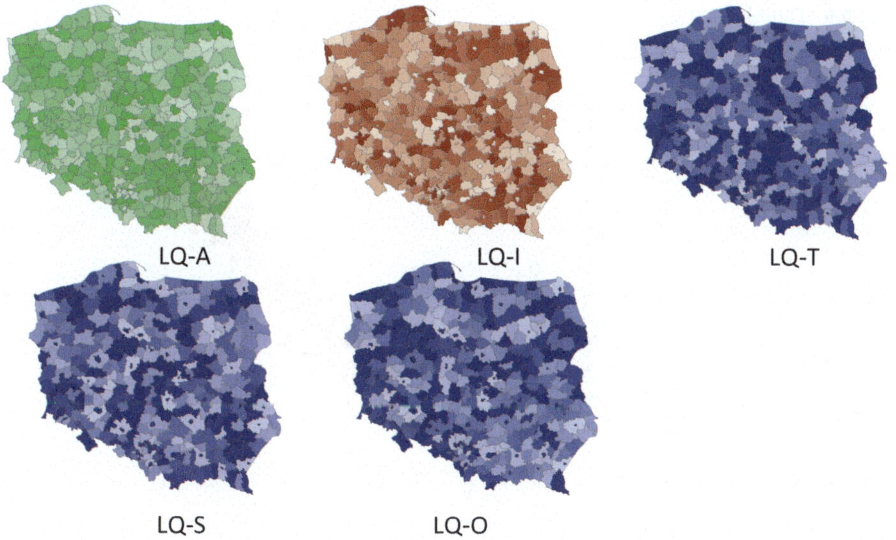

Fig. 6.4 SUREs for the employment categories

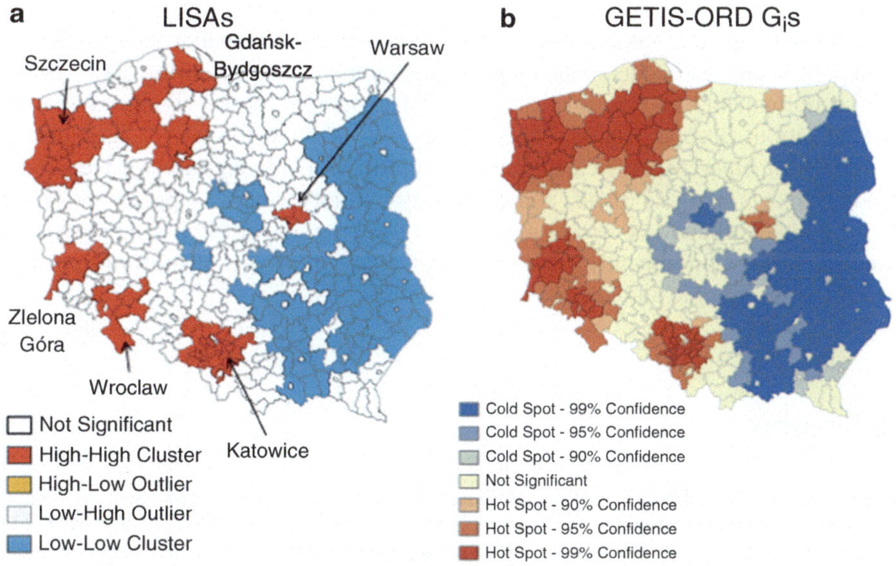

Fig. 6.5 LQ hot and cold spots. *Left* (**a**): LISAs. *Right* (**b**): Getis-Ord statistics

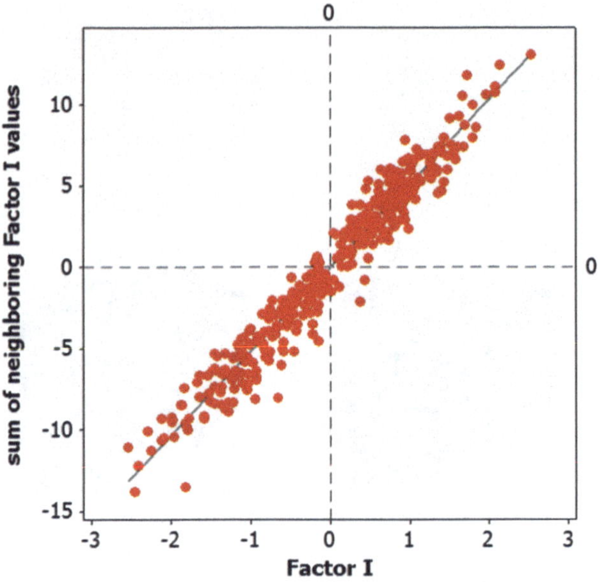

Fig. 6.6 The Moran scatterplot for the SSREs

6.8 Concluding Comments

A number of general conclusions can be drawn from this spatial analysis, even though little change occurs in the empirical geographic patterns during 2006–2013. First, employment LQs tend to cluster in Polish geographic space, signifying spatial specialization. Second, regional specialization is of three forms: (1) agricultural differs from other general categories of employment, (2) an east–west contrast gradient exists across Poland, and (3) a city-non-city contrast appears to exist. Third, spatial structure is very pronounced in the geography of Polish employment specialization. Consequently, this empirical analysis reveals that isomorphisms pertain to geographic and nongeographic clusters.

References

Beyene, J., & Moineddin, R. (2005). Methods for confidence interval estimation of a ratio parameter with application to location quotients. *BMC Medical Research Methodology, 5*, 32.

Carroll, M. C., Reid, N., & Smith, B. W. (2008). Location quotients versus spatial autocorrelation in identifying potential cluster regions. *The Annals of Regional Science, 42*(2), 449–463.

Cromley, R., & Hanink, D. (2012). Focal location quotients: Specification and applications. *Geographical Analysis, 44*, 398–410.

Feng, J., & Ji, M. (2011). Integrating location quotient, Local Moran's I and geographic linkage for spatial patterning of industries in Shanghai, China. In *Proceedings, 19th International*

Conference on Geoinformatics. Accessed October 15, 2017, from http://ieeexplore.ieee.org/document/5980854/
Griffith, D. (2004). A spatial filtering specification for the auto-logistic model. *Environment and Planning A, 36*, 1791–1811.
Moineddin, R., Beyene, J., & Boyle, E. (2003). On the location quotient confidence interval. *Geographical Analysis, 35*, 249–256.

Chapter 7
Spatial Autocorrelation Parameter Estimation for Massively Large Georeferenced Datasets

7.1 Introduction

In the early days of spatial autocorrelation (SA) analysis, datasets often were relatively small, with $n < 100$. Continual advancements in computer technology and more widespread georeferenced data collection resulted in n increasing, to 100s, then 1000s, and so forth, going beyond 10,000,000s today. Cressie et al. (1996) comment that such massively large georeferenced datasets challenge, and may ultimately cause the failure of, spatial statistical methodology initially designed for georeferenced datasets with n in the 100s, which has been extended successfully and effectively to datasets with n in the 10,000s. One example of this situation illustrated in this chapter is the implementation of spatial autoregressive models with remotely sensed satellite data. Another is the execution of simulation experiments, which is illustrated in this chapter with the Polish employment data. An important contribution of this chapter is the extension of the work reported in Griffith (2015).

7.2 Maximum Likelihood Estimation

The calculation of a maximum likelihood estimate (MLE) of the SA parameter in spatial regression furnishes one illustration of spatial statistical methodology developed for small-to-medium size samples that massively large samples defeat. Two features of this MLE are problematic: calculation of the Jacobian term (i.e., the normalizing constant ensuring that the complete set of probabilities integrates/sums to 1, which is the determinant of a positive definite n-by-n spatial covariance

Sections of this chapter are adapted from, and an extension of Griffith, D. (2015).

D. A. Griffith, J. H. P. Paelinck, *Morphisms for Quantitative Spatial Analysis*, Advanced Studies in Theoretical and Applied Econometrics 51,
https://doi.org/10.1007/978-3-319-72553-6_7

matrix) and minimization of the spatial regression nonlinear least squares objective function.

Griffith (2015) reviews solutions to the Jacobian problem, all of which are approximations. For a symmetric distribution of eigenvalues, such as that for a regular square tessellation, one Jacobian approximation is given by

$$-\sum_{i=1}^{n} \mathrm{LN}(1-\rho\lambda_j)/n \approx 2\,\omega\,\mathrm{LN}(\delta) - \omega\,\mathrm{LN}(\delta+\rho) - \omega\,\mathrm{LN}(\delta-\rho), \tag{7.1}$$

where the left-hand side is the exact Jacobian term, the right-hand side is its approximation, ρ is the spatial autoregressive parameter in a single-parameter spatial regression model specification, λ_j is the jth rank-ordered eigenvalue of the row-standardized spatial weights matrix (SWM) $\mathbf{W}$, and ω and δ are coefficients to be calibrated.[1] In addition, for a regular square lattice forming a complete rectangular region, the λ_j of the rook adjacency SWM $\mathbf{W}$ can be approximated by

$$-1 \le \lambda_{hk} = \mathrm{sign}\left\{\left|\left[\cos\left(\frac{h\pi}{P-1}\right) + \cos\left(\frac{k\pi}{Q-1}\right)\right]\Big/2\right|^{\gamma}\right\} \le 1, \tag{7.2}$$

where sign denotes the positive/negative sign of a value, and γ is an inflation exponent that asymptotically approaches 1 as n increases to infinity.

The spatial regression problem arises from the need to execute a nonlinear regression. The following is the likelihood function for a pure (i.e., it contains no covariates) SA SAR model specification:

$$L = (2\pi)^{-n/2}|\mathbf{V}|^{1/2}\left(\sigma^2\right)^{-n/2}\mathrm{e}^{-(\mathbf{Y}-\mu\mathbf{1})^{\mathrm{T}}\mathbf{V}(\mathbf{Y}-\mu\mathbf{1})/\left(2\sigma^2\right)}, \tag{7.3}$$

where $\mathbf{V}^{-1}\sigma^2 = [(\mathbf{I}-\rho\mathbf{W})^{\mathrm{T}}(\mathbf{I}-\rho\mathbf{W})]^{-1}\sigma^2$ is the spatial variance-covariance matrix, $|\mathbf{V}|$ denotes the determinant of matrix $\mathbf{V}$, $\mathbf{Y}$ is a n-by-1 vector of normally distributed random variable (RV), $\mathbf{1}$ is an n-by-1 vector of ones, T denotes the matrix transpose operation, and μ and σ^2, respectively, are the constant mean and variance of variable Y. MLEs for μ and σ^2 in Eq. (7.3) are

$$\widehat{\mu} = \mathbf{1}^{\mathrm{T}}\mathbf{V}\mathbf{Y}/\left(\mathbf{1}^{\mathrm{T}}\mathbf{V}\mathbf{1}\right), \tag{7.4}$$

and

$$\widehat{\sigma}^2 = (\mathbf{Y}-\mu\mathbf{1})^{\mathrm{T}}\mathbf{V}(\mathbf{Y}-\mu\mathbf{1})/n. \tag{7.5}$$

Simplification of the ρ estimation problem commonly has been achieved by substituting the two MLEs into Eq. (7.3), yielding

[1]Calibration refers to either estimation based upon a purposeful systematic sample covering a feasible parameter space or calculations based upon population moments.

$$L = (2\pi e/n)^{-n/2} |\mathbf{V}^{-1/n}|^{-n/2} \left[(\mathbf{Y} - \widehat{\mu}\mathbf{1})^{\mathrm{T}} \mathbf{V} (\mathbf{Y} - \widehat{\mu}\mathbf{1}) \right]^{-n/2}$$
$$= (2\pi e/n)^{-n/2} \left[|\mathbf{V}^{-1/n}| (\mathbf{Y} - \widehat{\mu}\mathbf{1})^{\mathrm{T}} \mathbf{V} (\mathbf{Y} - \widehat{\mu}\mathbf{1}) \right]^{-n/2}.$$

Because L is to be maximized, and the terms $|\mathbf{V}|$ and $\mathbf{V}$ containing ρ are raised to a negative exponent, the problem becomes one of

$$\text{MIN}: \quad |\,\mathbf{V}^{-1/n}\,|\,(\mathbf{Y} - \widehat{\mu}\mathbf{1})^{\mathrm{T}} \mathbf{V} (\mathbf{Y} - \widehat{\mu}\mathbf{1}), \tag{7.6}$$

which is equivalent to a weighted least squares regression problem. For a spatial SAR model specification, $|\mathbf{V}^{-1/n}|$ is equivalent to $e^{-\sum_{i=1}^{n} \mathrm{LN}(1-\rho\lambda_j)^2/n}$. Consequently, the constant mean MLE problem reduces to the following nonlinear least squares regression problem:

$$\frac{\mathbf{Y}}{e^{\sum_{i=1}^{n} \mathrm{LN}(1-\rho\lambda_j)/n}} = \frac{\rho\mathbf{W}\mathbf{Y}}{e^{\sum_{i=1}^{n} \mathrm{LN}(1-\rho\lambda_j)/n}} + \mu \frac{(\mathbf{I} - \rho\mathbf{W}\mathbf{Y})\mathbf{1}}{e^{\sum_{i=1}^{n} \mathrm{LN}(1-\rho\lambda_j)/n}} + \frac{\boldsymbol{\varepsilon}}{e^{\sum_{i=1}^{n} \mathrm{LN}(1-\rho\lambda_j)/n}}, \tag{7.7}$$

where ε is an independent and identically distributed (iid) Gaussian error term.

With massively large datasets, the challenge for Eq. (7.7), besides needing the n eigenvalues of the SWM, is that it requires nonlinear estimation that involves numerous iterations, each making many calculations with the full set of data. Employing Eqs. (7.1) and (7.2) to dramatically reduce the eigenvalues computational burden, the nonlinear regression computational burden can be reduced by resorting to the first partial derivative with respect to the SA parameter:

$$\frac{\partial LN(L)}{\partial \rho} = -\left(\sum_{j=1}^{n} \frac{\lambda_j}{1 - \rho\lambda_j} \right) / n$$
$$+ (\mathbf{Y} - \widehat{\mu}\mathbf{1})^{\mathrm{T}} (\mathbf{W}^{\mathrm{T}} + \mathbf{W} - \rho 2\mathbf{W}^{\mathrm{T}}\mathbf{W}) (\mathbf{Y} - \widehat{\mu}\mathbf{1}) / (2n\widehat{\sigma}^2) = 0. \tag{7.8}$$

Equation (7.8) involves many sums and cross-products that need to be calculated only once. Timing experiments reported in Griffith (2015) confirm replacing Eq. (7.7) with (7.8) results in a substantial reduction in computing time required to calculate $\widehat{\rho}$.

7.2.1 A Large Remotely Sensed Image Example

For illustrative purposes, a spatial autoregressive analysis was performed with a 4850-by-4988 pixels remotely sensed image of a highly polluted part of Peru[2]

[2]Mining began at the La Oroya, Peru, smelting site in 1893. Copper smelting began at the site in 1922 (70K tonnes capacity). Lead smelting began at the site in 1928 (120K tonnes capacity).

(Fig. 7.1a). The following measures, all of which are functions of spectral bands #3 and #4, were calculated with the remotely sensed image spectral data: ratio vegetation index (RVI), normalized difference vegetation index (NDVI), transformed soil adjusted vegetation index (TSAVI), optimized soil adjusted vegetation index (OSAVI), atmospherically resistant vegetation index (ARVI), brightness, greenness, wetness, and normalized burn ratio (NRB; Fig. 7.1b). Many of these indices are highly correlated; Table 7.1 summarizes their correlation matrix for this Peruvian empirical example:

Table 7.2 tabulates summary statistics for this example. This empirical illustration demonstrates that spatial autoregressive models can be estimated easily for massively large georeferenced datasets. It suggests that Eq. (7.2) performs better than Eq. (7.1)—consult $\widehat{\rho}$ for brightness, greenness, and wetness in Table 7.2—with this latter equation being applicable only for remotely sensed images. The nine selected spectral indices suggest that environmentally polluted areas of the world contain extremely high levels of positive SA [and hence considerable variance inflation (VIF)]. Figure 7.1b furnishes visual confirmation of this contention: it displays conspicuous map pattern and has $\widehat{\rho} = 0.96$. This illustration also highlights how statistical significance becomes compromised with massively large sample sizes: virtually everything is statistically significant when the magnitude of standard errors is of the order 0.0001.

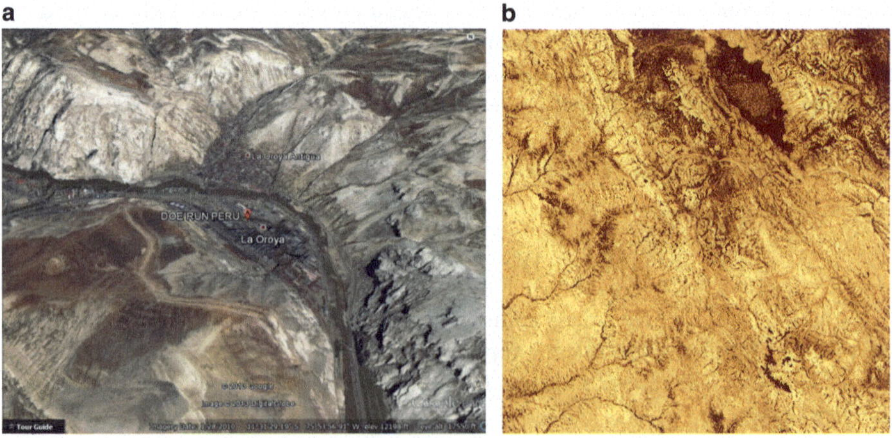

Fig. 7.1 La Oroya, Peru smelting site. Left (**a**): site of the smelter and its surroundings. Right (**b**): the geographic distribution of the NRB index for the site. These are Landsat 4-5 TM C! Level-1 images of the La Oroya, Peru area for 20 June 1997, downloaded from the webpage http://glovis.usgs.gov/ (last accessed on 3 November 2017); the visualization in (**a**) is a composite image constructed by the author using all spectral bands

---

Finally, zinc smelting began at the site in 1952 (45K tonnes capacity). Air pollution readings in 1999 indicated extreme levels of arsenic, cadmium, and lead.

Table 7.1 Product moment correlation matrix for the Fig. 7.1a remotely sensed image spectral indices

Spectral index	RVI	NDVI	OSAVI	ARVI	TSAVI	NBR	Brightness	Greenness	Wetness
RVI	1.00								
NDVI	0.96	1.00							
OSAVI	0.96	1.00	1.00						
ARVI	0.44	0.46	0.46	1.00					
TSAVI	0.95	1.00	1.00	0.48	1.00				
NBR	0.63	0.54	0.54	0.18	0.51	1.00			
Brightness	0.16	0.21	0.21	−0.10	0.27	−0.11	1.00		
Greenness	0.72	0.68	0.68	0.31	0.64	0.62	−0.47	1.00	
Wetness	−0.22	−0.33	−0.33	0.08	−0.37	0.37	−0.79	0.24	1.00

RVI, NDVI, and TSAVI essentially are the same. NBR is a useful index for detecting fire burn

Table 7.2 Selected spectral index SAR results for the La Oroya, Peru, smelter site

Index	Jacobian equation	$\widehat{\mu}$	$\widehat{\sigma}_{\widehat{\mu}}$	$\widehat{\sigma}^2$	Variance inflation	$\widehat{\rho}$	$\widehat{\sigma}_{\widehat{\rho}}$
RVI	(7.1)	1.3231	0.0015	0.0068		0.9891	0.0001
	(7.2)	1.3245		0.0069	28.2	0.9805	
NDVI	(7.1)	0.1135	0.0039	0.0007		0.9865	0.0001
	(7.2)	0.1136		0.0007	29.4	0.9787	
OSAVI	(7.1)	0.1314	0.0005	0.0009		0.9866	0.0001
	(7.2)	0.1316		0.0009	29.5	0.9788	
ARVI	(7.1)	0.4579	0.0003	0.0104		0.9224	0.0001
	(7.2)	0.4583		0.0104	6.6	0.9209	
TSAVI	(7.1)	0.1327	0.0004	0.0007		0.9873	0.0001
	(7.2)	0.1329		0.0007	29.7	0.9794	
NBR	(7.1)	0.0204	0.0002	0.0020		0.9615	0.0001
	(7.2)	0.0204		0.0020	13.5	0.9543	
Brightness	(7.1)	95.4732	9.8263	23.3875		0.9999	0.0001
	(7.2)	95.6072		23.4812	44.2	0.9987	
Greenness	(7.1)	−24.6370	3.8187	3.5320		0.9999	0.0001
	(7.2)	−24.6719		3.5841	48.4	0.9933	
Wetness	(7.1)	−56.9633	9.4764	21.7516		0.9999	0.0001
	(7.2)	−57.0196		22.0053	42.3	0.9951	

7.2.2 *Other Approaches*

Walde et al. (2008) address the issue of computing MLEs for massively large georeferenced datasets, acknowledging the challenge of calculating Jacobians and maintaining numerical accuracy. They dislike the Jacobian approximation approach utilized in the preceding section. They endorse the generalized method of moments (GMM) approach of Kelejian and Prucha (e.g., 2010), which is a popular estimation technique in spatial econometrics. They argue that its advantages include a directly computable standard error and avoidance of a grid search. But their analyses are based upon a 1-NN SWM.

Burden et al. (2015) also acknowledge the computational complications that arise with estimating parameters of an SAR model, presenting a methodology that relates to eigenvector spatial filtering. They capture SA with a spatially structured random effects (SSRE) term (see Sect. 4.2), describing the SA with an SAR model specification.

7.3 The Sampling Variance of $\widehat{\rho}$

Calculating $\widehat{\rho}$ should be accompanied with a measure of uncertainty (i.e., its standard error). The most common null hypothesis is that georeferenced data contain no SA (i.e., H_0: $\rho = 0$). Ord (1975) furnishes the inverse asymptotic variance matrix for the SAR model specification. Applying the appropriate matrix algebra operations yields

$$\text{ASY}\sigma^2_{\widehat{\rho}} = \frac{1}{\text{TR}\left[(\mathbf{I}-\rho\mathbf{W}^{\text{T}})^{-1}\mathbf{W}^{\text{T}}\mathbf{W}(\mathbf{I}-\rho\mathbf{W})^{-1}\right]+\sum_{j=1}^{n}\frac{\lambda_j^2}{(1-\rho\lambda_j)^2}-\frac{2}{n}\left\{\text{TR}\left[\mathbf{W}(\mathbf{I}-\rho\mathbf{W})^{-1}\right]\right\}^2} \tag{7.9}$$

where subscript ASY denotes asymptotic, TR denotes the matrix trace operation, superscript T denotes the matrix transpose operation, and **I** is an n-by-n identity matrix.

If $\rho = 0$, Eq. (7.9) reduces to

$$\text{ASY}\sigma^2_{\widehat{\rho}} = \frac{1}{\text{TR}(\mathbf{W}^{\text{T}}\mathbf{W}) + \sum_{j=1}^{n}\lambda_j^2}, \tag{7.10}$$

given that TR(**W**) = 0, where, for a P-by-Q regular square tessellation forming a complete rectangular region and a rook adjacency SWM, $\text{TR}(\mathbf{W}^{\text{T}}\mathbf{W}) = \sum_{j=1}^{n}\lambda_j^2 + \frac{P+Q+12}{72}$. This last result derives from TR($\mathbf{W}^{\text{T}}\mathbf{W}$) = TR ($\mathbf{C}^2\mathbf{D}^{-2}$) $= \sum_{i=1}^{n}\frac{1}{\sum_{j=1}^{n}c_{ij}}$, where **D** is a diagonal matrix whose ith diagonal entry is $\sum_{j=1}^{n}c_{ij}$. This quantity is relatively easy to calculate; it is the sum of the inverse of the SWM row totals.[3] For a P-by-Q regular square tessellation and a rook adjacency SWM, it equals

$$(3PQ + 2P + 2Q + 4)/12,$$

whereas $\sum_{j=1}^{n}\lambda_j^2$ equals

[3]For a P-by-Q regular square tessellation and a queen adjacency SWM, $\text{TR}(\mathbf{W}^{\text{T}}\mathbf{W}) = \sum_{j=1}^{n}\lambda_j^2 + \frac{81P+81Q+326}{2400}$. For a P-by-Q (horizontal-by-vertical) regular hexagonal tessellation SWM, $\text{TR}(\mathbf{W}^{\text{T}}\mathbf{W}) = \sum_{j=1}^{n}\lambda_j^2 + \frac{5P+12Q+25}{180}$. The limit of these correction factors goes to zero as the size of a surface partitioning goes to infinity:

$$(18PQ + 11P + 11Q + 12)/72,$$

The difference between these two quantities is $(P + Q + 12)/72$. Therefore, for a P-by-Q regular square tessellation, a rook adjacency SWM, and $\rho = 0$,

$$_{\mathrm{ASY}}\sigma^2_{\widehat{\rho}} = \frac{72}{36PQ + 23P + 23Q + 36}. \tag{7.11}$$

Two noteworthy features here are: (1) $\sigma\widehat{\rho}^2$ is a function of only the number of 1s in a SWM; and, (2) as the size of a surface partitioning increases to infinity, the standard error of $\widehat{\rho}$ converges on 0.

Because $\mathrm{TR}\left(\mathbf{W}^{\mathrm{T}}\mathbf{W}\right) = \sum_{j=1}^{n} \lambda_j^2$ when matrix $\mathbf{W}$ is symmetric (Ord 1975), the following is a good approximation to $_{ASY}\sigma^{ASY}\widehat{\rho}^2$ for $\rho = 0$:

$$_{\mathrm{ASY}}\sigma^2_{\widehat{\rho}} = \frac{1}{2\sum_{j=1}^{n}\lambda_j^2} = \frac{72}{36PQ + 22P + 22Q + 24}. \tag{7.12}$$

The limit of the ratio of Eqs. (7.11) and (7.12) as a surface partitioning goes to infinity is 0. For $n = 100$, Eq. (7.12) underestimates the asymptotic variance by less than 0.8%; for $n = 313{,}853$ (the size of some smaller remotely sensed images), this underestimation is roughly 0.01%; for $n = 30{,}916{,}600$ (the size of some larger remotely sensed images), this underestimation is roughly 0.0001%.

Consequently, an approximation to Eq. (7.9) for $\rho \neq 0$ meriting assessment is

$$_{\mathrm{ASY}}\sigma^2_{\widehat{\rho}} = \frac{1}{2\sum_{j=1}^{n}\frac{\lambda_j^2}{\left(1-\rho\lambda_j\right)^2} - \frac{2}{n}\left[\sum_{j=1}^{n}\frac{\lambda_j}{\left(1-\rho\lambda_j\right)}\right]^2}. \tag{7.13}$$

The right-hand term in the denominator of Eq. (7.9) can be simplified to the right-hand term appearing in Eq. (7.13).

Theorem 7.1

$$\mathrm{TR}\left[\mathbf{W}(\mathbf{I} - \rho\mathbf{W})^{-1}\right] = \sum_{j=1}^{n}\frac{\lambda_j}{(1 - \rho\lambda_j)}.$$

PF

$$\mathrm{TR}\left[\mathbf{W}(\mathbf{I} - \rho\mathbf{W})^{-1}\right] = \mathrm{TR}\left[(\mathbf{I} - \rho\mathbf{W})^{-1}\mathbf{W}\right]$$

$$= \mathrm{TR}\left[\left(\mathbf{I} - \rho\mathbf{D}^{-1}\mathbf{C}\right)^{-1}\mathbf{D}^{-1}\mathbf{C}\right]$$

$$= \mathrm{TR}\left\{\left[\mathbf{D}^{-1/2}\left(\mathbf{I} - \rho\mathbf{D}^{-1/2}\mathbf{C}\mathbf{D}^{-1/2}\right)\mathbf{D}^{1/2}\right]^{-1}\mathbf{D}^{-1}\mathbf{C}\right\}$$

$$= \mathrm{TR}\left[\left(\mathbf{I} - \rho\mathbf{D}^{-1/2}\mathbf{C}\mathbf{D}^{-1/2}\right)^{-1}\mathbf{D}^{-1/2}\mathbf{C}\mathbf{D}^{-1/2}\right]$$

$\mathbf{D}^{-1/2}\mathbf{C}\mathbf{D}^{-1/2}$ is a symmetric matrix whose eigenvalues are the same as those for matrix **W**

$$\therefore \quad \mathrm{TR}\left[\mathbf{W}(\mathbf{I} - \rho\mathbf{W})^{-1}\right] = \sum_{j=1}^{n} \frac{\lambda_j}{(1 - \rho\lambda_j)}$$

A correction factor[4] that is a function of both ρ (Fig. 7.2a) and n needs to be added to the denominator of Eq. (7.13). Nevertheless, even for $n = 20^2$, the relative magnitude of this correction factor ranges from about 0.7% to 1.5% across the values of ρ, for a rook adjacency SWM; it is not substantial. Meanwhile, Eq. (7.2) can be used to approximate the eigenvalues for all necessary calculations, estimating its exponent γ such that $\sum_{j=1}^{n} \widehat{\lambda}_j^2 = (18PQ + 11P + 11Q + 12)/72$, where $\widehat{\lambda}_j$ $(j = 1, 2, \ldots, n)$ denotes the approximated eigenvalues.

A comparison here of the SAR asymptotic and relatively small sample ($n = 10^2$ to 75^2) variances for a rook adjacency SWM with $\rho = 0$ (the frequency distribution of $\widehat{\rho}$ conforms to a bell shaped curve in each of the 14 cases) is based upon a simulation experiment with 10,000 replications. Figure 7.2b portrays the scatterplot of simulated and asymptotic standard errors (versus the square root of n for a $P = Q$ regular square tessellation) from this simulation experiment: both are converging on 0, and the asymptotic standard error is converging on the actual (i.e., simulated) standard error. Of concern is that the asymptotic standard error is nearly 50% greater than the actual standard error when n is 100, and remains sizeable for sample sizes as large as 1000.

7.3.1 The Asymptotic Variance for Massively Large Georeferenced Datasets: The First-Order Eigenvalue Term

Remotely sensed images are one source of massively large georeferenced datasets. Their SWM often is the rook adjacency for a regular square tessellation forming a complete rectangular region. Equation (7.2) furnishes an eigenvalue approximation for their SWM **W**. Because the set of eigenvalues falls into the interval $[-1, 1]$, as n increases, the rank ordered set of eigenvalues divides this interval into

[4]For a given n, this correction factor appears to be of the form $\frac{P+Q+12}{72} - \beta\left(\frac{2}{\delta}\right)^{\gamma} + \alpha + \beta\left(\frac{1}{\delta+\rho} + \frac{1}{\delta-\rho}\right)^{\gamma}$. For $n = 20^2$, $k = 0.7222$, $\widehat{\alpha} = -0.0872$, $\widehat{\beta} = 0.4414$, $\widehat{\delta} = 1.0011$, $\widehat{\gamma} = 1.4571$, and the RESS $= 1.1 \times 10^{-5}$.

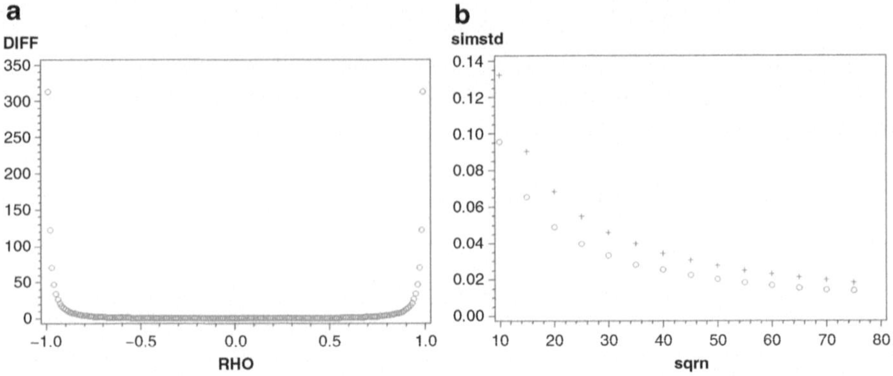

Fig. 7.2 Left (**a**): A correction factor for Eq. (7.13) for a rook adjacency SWM and $n = 20^2$. Right (**b**): comparison of asymptotic (open circles) and simulated (pluses) standard errors for a range of n

increasingly smaller and smaller segments. As such, it is reminiscent of the trapezoid rule from calculus that furnishes an approximation for a definite integral; here the integral is $\int_{-1}^{1} \lambda \mathrm{d}\lambda$, where λ denotes the eigenvalue variable. One modification required here arises from the sum of the eigenvalues referred to in Theorem 7.1 not being identical to the summation in the trapezoid rule. Therefore, the equivalency is given by, for $\rho = 0$,

$$\sum_{j=1}^{n} \lambda_j \approx \frac{n}{2}\left[\int_{-1}^{1} \lambda \mathrm{d}\lambda + \frac{1}{n}(-1+1)\right].$$

This approximation is an equality; both sides of the approximation sign equal 0. But Theorem 7.1 concerns $\sum_{j=1}^{n} \frac{\lambda_j}{(1-\rho\lambda_j)}$ when $\rho \neq 0$, too. Now the approximation is

$$\sum_{j=1}^{n} \frac{\lambda_j}{(1-\rho\lambda_j)} \approx \frac{n}{2}\left\{\int_{-1}^{1} \frac{\lambda}{1-\rho\lambda}\mathrm{d}\lambda + \frac{1}{n}\left[\frac{-1}{1+\rho} + \frac{1}{1-\rho}\right]\right\},$$

which contains a conspicuous small n bias as ρ deviates from 0. The corrected approximation is given by

$$\left[\frac{n(1-\rho^2)\text{TANH}^{-1}(\rho)+(n+1)\,\rho^3-n\,\rho}{\rho^2\,(1-\rho^2)}\right]^{\alpha+\beta|\rho|}, \tag{7.14}$$

where $\widehat{\alpha}=1-\frac{0.17674}{(\sqrt{n}-4.6533)^{0.32982}}$, and $\widehat{\beta}=\frac{1.33708}{(\sqrt{n}+1.84535)^{1.20333}}$. The first of these exponent terms converges on 1, and the second converges on zero, as n goes to infinity. For n = 40,000,000 (a rather standard size remotely sensed image), $\widehat{\alpha}=0.99010$ and $\widehat{\beta}=0.00004$.

Similar results need to be established for other SWMs.

7.3.2 The Asymptotic Variance for Massively Large Georeferenced Datasets: The Second-Order Eigenvalue Term

Considering the trapezoid rule again, here the variable of interest is λ^2:

$$\sum_{j=1}^{n}\lambda_j^2\approx\frac{n}{2}\left[\int_{-1}^{1}\lambda^2\text{d}\lambda+\frac{1}{n}(1+1)\right]\Rightarrow\frac{n}{4}+\frac{11\sqrt{n}}{36}+\frac{1}{6}\approx\frac{n}{3}+1.$$

This approximation is not good; asymptotically, the ratio of the trapezoid approximation to the actual sum of squared eigenvalues is 4/3, implying that the trapezoid approximation requires some type of correction transformation. One source of this deviation is the failure of eigenvalues to partition the interval [−1, 1] into uniform segments, which the trapezoid rule assumes. This term has the form portrayed in Fig. 7.2a.

The approximation of interest here may be written as

$$\sum_{j=1}^{n}\frac{\lambda_j^2}{(1-\rho\lambda_j)^2}\approx\alpha+\beta\,\text{I}_{\text{sign}}\left|\frac{n}{2}\left\{\int_{-1}^{1}\frac{\lambda^2}{(1-\rho\lambda)^2}\text{d}\lambda+\frac{1}{n}\left[\frac{1}{(1+\rho)^2}+\frac{1}{(1-\rho)^2}\right]\right\}\right|^{\gamma},$$

where I_{sign} denotes the positive/negative sign of the quantity within the absolute value signs.[5] The right-hand side of this approximation equals

$$\alpha+\beta\text{I}_{\text{sign}}\left|\frac{\rho^5+\rho^3+n\,\rho(1-\rho^2)(2-\rho^2)-2n(1-\rho^2)^2\text{TANH}^{-1}(\rho)}{\rho^3(1-\rho^2)^2}\right|^{\gamma}, \tag{7.15}$$

which is discontinuous at $\rho=0$, the value of ρ for which the expression already has been established to be $n/3+1$. Numerical evaluation of rook adjacency regular square lattice cases for $P=Q=10$ to $P=Q=80$ suggests that

[5]This formulation allows negative values to be raised to noninteger exponents.

$$\widehat{\alpha} = 45.71886 \left(\sqrt{n}\right)^{0.85561} \left[1 - e^{-0.00121(\sqrt{n})^{0.85561}}\right],$$

$$\widehat{\beta} = 1.04913 - 0.595116e^{-0.03842(\sqrt{n})^{0.66248}}, \text{ and}$$

$$\widehat{\gamma} = 1 + \frac{2.91148}{\sqrt{n} + 10} - \frac{0.16782}{\sqrt{n} - 3}.$$

Although they require considerably more refinement, each of these estimates is accompanied by a very small relative error sum of squares. Furthermore, they render a correct $\rho = 0$ limit:

$$\operatorname*{limit}_{n \to \infty} \frac{\alpha + \beta \, \mathrm{I_{sign}} \left| \frac{\rho^5 + \rho^3 + n\,\rho\left(1-\rho^2\right)\left(2-\rho^2\right) - 2n\left(1-\rho^2\right)^2 \mathrm{TANH}^{-1}(\rho)}{\rho^3\left(1-\rho^2\right)^2} \right|^{\gamma}}{(18n + 22\sqrt{n} + 12)/72} = 1.$$

Consequently, expression (7.15) should prove very useful for spatial regression analyses of massively large remotely sensed images.

7.3.3 *The Asymptotic Variance for Massively Large Georeferenced Datasets: The Residual Term*

When the SWM **W** is asymmetric, the term TR[$(\mathbf{I} - \rho\mathbf{W}^{\mathrm{T}})^{-1}\mathbf{W}^{\mathrm{T}}\mathbf{W}(\mathbf{I} - \rho\mathbf{W})^{-1}$] in the asymptotic variance comprises two terms, as outlined in Sect. 3: $\sum_{j=1}^{n} \frac{\lambda_j^2}{\left(1 - \rho\lambda_j\right)^2}$ and an expression that reduces to $(P + Q + 12)/72$ for a rook adjacency regular square tessellation and $\rho = 0$. Approximating this second term, which has the form portrayed in Fig. 7.2a, is the subject of this section.

Based upon matrix trance results from $P = Q = 10$ to 80, this term may be described by the following expression:

$$\frac{\sqrt{n} + 6}{36} - 2\beta + \frac{\beta}{(1 + \rho)^{\gamma}} + \frac{\beta}{(1 - \rho)^{\gamma}},$$

where $\widehat{\beta} = 0.06696 + 0.01659\sqrt{n}$, and $\widehat{\gamma} = 1.47825 + e^{-1.85423(\sqrt{n})^{0.22698}}$. This term is relatively small compared with $\sum_{j=1}^{n} \frac{\lambda_j^2}{\left(1 - \rho\lambda_j\right)^2}$.

7.3.4 A Preliminary Asymptotic Variance Approximation Accuracy Assessment

By definition, the approximations are not exact. Figure 7.2b also confirms that the asymptotic variance—Eq. (7.13)—also is not exact. Table 7.3 tabulates selected comparisons of these variances, for a standard null hypothesis and for a typical remotely sensed image case, in terms of their square roots (i.e., standard errors).

For $\rho = 0$, $_{\text{ASY}}\sigma^2_{\widehat{\rho}}$ values result from Eq. (7.12); for $\rho = 0.9$ $_{\text{ASY}}\sigma^2_{\widehat{\rho}}$ values result from numerical matrix calculations.

These findings indicate that more work needs to be undertaken to improve the approximation and to extend it to combinations of P and Q other than $P = Q$. Nevertheless, it does furnish a good result for massively large datasets (e.g., $10{,}000^2 = 100{,}000{,}000$, which is greater than the number of pixels in most current remotely sensed images).

7.4 Irregular Surface Partitioning Spatial Autocorrelation Simulation Experiments

Figure 7.2b rests on results from a simulation experiment (as do the analyses in Sects. 2.5 and 5.2). Rather than a massively large cross-sectional dataset, such as a remotely sensed image, a simulation experiment with 10,000 replications (to exploit the Law of Large Numbers) of n areal unit values also can constitute a massively large georeferenced dataset.

This section summarizes findings about the sample variation of $\widehat{\rho}$ for a georeferenced binomial RV, using the 2013 Polish location quotient (LQ) data for the industrial employment category. An eigenvector spatial filter (ESF) description of these binomial data includes 23 of 85 candidate eigenvectors, accounts for roughly 50% of their geographic variability, and renders a bivariate intercept term of 0.00226 and a bivariate slope coefficient of 0.99190. In contrast, following a logit specification, the pure spatial autoregressive specification estimated here is

Table 7.3 Selected remotely sensed image spatial autoregressive parameter standard errors for the null and a particular alternate hypothesis

	$\rho = 0$		$\rho = 0.9$	
n	$_{\text{ASY}}\sigma^2_{\widehat{\rho}}$	$\widehat{\sigma}^2$	$_{\text{ASY}}\sigma^2_{\widehat{\rho}}$	$\widehat{\sigma}^2$
10^2	0.13258	0.13376	0.03862	0.03957
30^2	0.04614	0.04210	0.01468	0.01319
70^2	0.02002	0.01773	0.00654	0.00570
1000^2	0.00141	0.00135	***	0.00045
$10{,}000^2$	0.00014	0.00014	***	0.00005

Notes: *** denotes unknown

$$p_i = \frac{1}{1 + \exp\left\{-\left[\alpha + \rho \sum_{j=1}^{n} w_{ij} \mathrm{LN}\left(\frac{p_j}{1-p_j}\right)\right]\right\}},$$

where exp denotes the antinatural logarithm, and p_i is the numerator of the LQ for powiat i. This autoregressive description of a binomial RV is less successful than its ESF counterpart: it accounts for roughly 1/3 of the geographic variance in the LQ under study and has a bivariate intercept term of −0.00329 and a bivariate slope coefficient of 1.01180. The spatial autocorrelation parameter estimate is $\widehat{\rho} = 0.6676$ (se = 0.0242).

A simulation experiment was executed using the observed probabilities to draw 10,000 random samples from a binomial distribution with the correct mean and 4000 trials[6] (i.e., N_t); upon completion of this experiment, the computed variance was rescaled for 100 trials (it equals $0.30/\sqrt{N_t}$). This is a massive georeferenced dataset because it involves 3,800,000 areal unit values, each being sampled from a distribution with 4000 values. The average spatial autocorrelation parameter estimate is $\overline{\widehat{\rho}} = 0.6660$; the empirical estimate of this parameter is unbiased. However, the simulated standard error is 0.03281, which differs from the reported standard error of 0.0242; spatial autocorrelation mostly impacts upon variances and hence standard errors. Because parameter estimates in this experiment resemble pseudo-likelihood ones, a proper analysis requires execution of Markov chain Monte Carlo (MCMC) maximum likelihood estimation, which would increase the data handling size by a factor of as much as 100,000.

7.5 Concluding Comments

This chapter highlights a number of important spatial statistics concepts and topics that pertain to both regular and irregular surface partitionings. Foremost is that maximum likelihood techniques can be used with massively large datasets, particularly remotely sensed images. Furthermore, conceptualizations exist that allow standard errors to be either computed or approximated without having to resort to the calculation of eigenfunctions or inverses of massively large matrices, computations that become impossible when n reaches a certain elusive gigantic number. Cressie et al. (1996) comment that massively large georeferenced datasets challenge, and may ultimately cause the failure of, conventional spatial statistical methodology. Many of the standard errors discussed in this chapter are close to zero when n is massively large, which means virtually any result becomes statistically significant. This outcome certainly is a failure that needs to be rectified: what

[6]This value minimized the chances of having a count of 0 or100, neither of which occurs in the empirical data.

is meant by uncertainty and statistical importance when a map is divided into a massively large number of areal units?

Findings summarized in this chapter are at odds with Walde et al.'s (2008) contention about Ord's (1975) approach to estimating linear spatial autoregression model parameters. More work needs to be devoted to their types of comparison. In addition, applied spatial statistics needs to better embrace GMM estimates as an alternative to MLEs.

References

Burden, S., Cressie, N., & Steel, D. (2015). The SAR model for very large datasets: A reduced rank approach. *Econometrics, 3*, 317–338.

Cressie, N., Olsen, A., & Cook, D. (1996). Massive data sets: Problems and possibilities, with application to environmental monitoring. In The Committee of Applied and Theoretical Statistics (Ed.), *Massive data sets: Proceedings of a Workshop* (pp. 115–119). Washington, DC: National Academy Press.

Griffith, D. (2015). Approximation of Gaussian spatial autoregressive models for massive regular square tessellation data. *International Journal of Geographical Information Science, 29*, 2143–2173.

Kelejian, H., & Prucha, I. (2010). Specification and estimation of spatial autoregressive models with autoregressive and heteroskedastic disturbances. *Journal of Econometrics, 157*, 53–67.

Ord, J. (1975). Estimation methods for models of spatial interactions. *Journal of the American Statistical Association, 70*, 120–126.

Walde, J., Larch, M., & Tappeiner, G. (2008). Performance contest between MLE and GMM for huge spatial autoregressive models. *Journal of Statistical Computation and Simulation, 78*, 151–166.

Chapter 8
Space–Time Data and Semi-saturated Fixed Effects

8.1 Introduction

Section 3.5 mentions fixed effects (FE), whereas Sect. 4.2 discusses random effects (RE). The goal of these specifications is to remove omitted variable bias (OVB) by measuring change over time within individual areal units. The topic of this chapter is FE, whose meaning is better understood in the context of RE (Bell and Jones 2015). These two concepts relate to both spatial and temporal autocorrelation, heterogeneity, and time invariant components of space–time data. A FE specification assumes that areal unit specific effects are, whereas a RE specification assumes that they are not, correlated with covariates. If the FE assumption prevails, then a FE specification yields more efficient estimators, and a RE specification yields inconsistent estimators; if it does not, then it yields less efficient estimators. A FE specification can be implemented with binary 0-1 indicator (i.e., dummy, categorical) variables. An indicator variable is defined for each point in time and for each areal unit in the cross-sectional part of space–time dataset. The sets of indicator variables allow the calculation of global, temporal, and areal unit specific means. This conceptualization is similar to a two-way analysis of variance (ANOVA) experimental design for which the two factors are time and geography. But ANOVA has multiple observations in each cross-classification, permitting analysis of time, areal unit, and interaction terms. A space–time dataset has only one observation in each cross-classification, permitting analysis of only time and areal unit, but not interaction, terms.

Because T indicator variables are used for a time series, and n indicator variables are used for the areal units, each of these two sets of indicator variables sums to a vector of ones, which is the covariate for an intercept term in a regression model specification. Consequently, an arbitrarily selected indicator variable from each set needs to be set aside (i.e., its coefficient is equated to 0) in order to avoid perfect multicollinearity. If these selected indicator variables are excluded from a

D. A. Griffith, J. H. P. Paelinck, *Morphisms for Quantitative Spatial Analysis*,
Advanced Studies in Theoretical and Applied Econometrics 51,
https://doi.org/10.1007/978-3-319-72553-6_8

specification, then their means equal the intercept term, and the remaining indicator variable coefficients are added to the intercept term. If these selected indicator variables are subtracted from each of the others in their respective sets, then the resulting regression coefficients are differences of conditional means.

Meanwhile, a saturated FE specification contains the same number of parameters as observations (i.e., an over-parameterized specification). For space–time data, the interaction terms in the T-by-n cross-classification conform to this property. Semi-saturated may be defined as follows: the indicator variables indexing these interaction terms are irrelevant, and the only relevant indicator variables are for the time- and areal unit-specific factors. This is the specification employed in Sect. 3.5 that is labeled approximately, not fully, saturated.

8.2 What Is Fixed Effects?

A FE approach to space–time data analysis supports the approximation of randomized experiments by quasi-experiments. A FE specification links to the values in a space–time dataset in terms of the aforementioned indicator variables, which are nonrandom: time points do not change, and areal units do not change. These indicator variables yield time-specific and areal unit-specific means. Respectively, these individual means are treated as being invariant (i.e., fixed) over geography and over time. The FE term controls for observable and unobservable covariates across areal units, removing between variation and leaving within variation (in ANOVA parlance), eliminating OVB when the effects are time invariant. However, if the effects are not time invariant (i.e., not fixed), then the possibility of OVB remains. Furthermore, because FE completely ignores between areal unit variation, its inferential basis relates to increased sampling variability and hence larger standard errors.

Inference is one primary difference between FE and RE specifications. FE supports inference about only the set of n areal units and their observed response variable values and not beyond the analyzed dataset. In contrast, a RE specification supports inference about the parent population from which the set of areal units and their response variable values were drawn. Another difference between these two specifications is the number of parameters: the FE specification has K covariate coefficients plus $n-1$ areal unit coefficients (as well as $T-1$ time coefficients), whereas the RE specification has K covariate coefficients plus two distributional parameters. The FE specification also tends to be more robust, although the indicator variables preclude inclusion of time invariant regressors, and is simpler to estimate, but can exacerbate measurement error bias.

A FE specification has two data requirements: (1) each areal unit must have at least two response variable measurements, and (2) at least some of these repeated measurement values must be different. In general, the first requirement is additional information, which also can be supplied by a Bayesian specification (see Sect. 4.3).

This areal unit specific conceptualization can be extended to covariate regression coefficients, too. This is an analysis of covariance (ANCOVA) specification.

8.3 Testing for Fixed Effects

The Durbin-Wu-Hausman (DWH; Nakamura and Nakamura 1981) statistic furnishes the test of choice for differentiating between a FE and a RE specification. As noted in the preceding section, one fundamental difference between these two specifications is that FE yields consistent, whereas RE yields inconsistent, estimators. The EWH test exploits this difference as well as the FE specification being more efficient when the RE assumption is invalid. Accordingly, estimates from the two estimators should be similar when both are consistent and dissimilar when the RE assumption is invalid. The DHW test results in a failure to reject the null hypothesis of RE when the coefficients are similar.

The DWH test statistic is given by

$$(\mathbf{b}_{\mathrm{FE}} - \mathbf{b}_{\mathrm{RE}})^{\mathrm{T}}[\mathrm{VAR}(\mathbf{b}_{\mathrm{FE}}) - \mathrm{VAR}(\mathbf{b}_{\mathrm{RE}})]^{-1}(\mathbf{b}_{\mathrm{FE}} - \mathbf{b}_{\mathrm{RE}}), \tag{8.1}$$

where K is the number of covariates (excluding the intercept), $\mathbf{b}_{\mathrm{FE}}$ is the K-by-1 vector of FE covariate regression coefficients, $\mathbf{b}_{\mathrm{RE}}$ is the K-by-1 vector of RE covariate regression coefficients, and VAR denotes the K-by-K variance-covariance matrix of its regression coefficient argument. DWH is distributed as a chi-squared random variable with K degrees of freedom. Expression (8.1) becomes problematic when it is negative.

Clarke et al. (2010) argue that for situations in which the assignment of phenomena to areal units is poorly understood, the robustness of FE estimators outweighs any attractive features of the RE specification. Accordingly, the FE specification often should be preferable to the RE specification.

8.4 Fixed Effects: SSFE and SUFE Components

The spatially structured fixed effects (SSEF) tend to be nearly identical to the spatially structured random effects (SSRE), as do the spatially unstructured fixed effects (SUEF) and spatially unstructured random effects (SURE), for the categorized Polish employment data (e.g., Fig. 8.1, which parallels Sect. 4.2 results).

In general, both of the SSFE-SSRE and SUSF-SURE correspondences are reasonably good (Table 8.1). However, some conspicuous as well as inconsequential discrepancies exist in the five categorized Polish employment cases. For $LQ_{\mathrm{agricullture}}$, the SSFE contains a single extreme outlier missing from the SSRE; Table 8.1 calculation results are for after its removal. For LQ_{finance}, the SSFE contains three extreme outliers missing from the SURE; again, Table 8.1

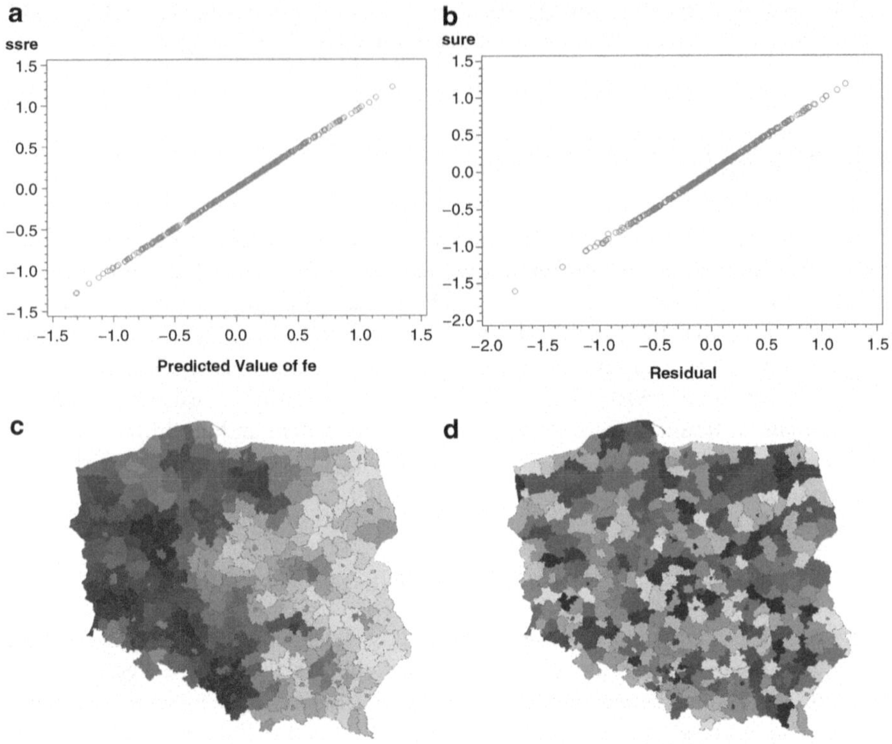

Fig. 8.1 $LQ_{industry}$ scatterplots. Top left (**a**): SSRE and SSFE. Top right (**b**): SURE and SUFE. Choropleth map geographic distributions: the darkness of the grayscale is directly proportional to the magnitude of the value. Bottom left (**c**): SSFE (MC = 0.96, GR = 0.21). Bottom right (**d**): SUFE (MC = −0.06, GR = 1.01)

Table 8.1 Polish employment categories' bivariate regression results for the FE and RE terms

Employment category	SSFE			SURE		
	Intercept	Slope	R^2	Intercept	Slope	R^2
A	0	0.9668	0.9997	0	0.9672	0.9992
I	0	0.9711	0.9999	0	0.9676	0.9995
T	0	0.9428	0.9997	0	0.9492	0.9989
F	0	0.7451	0.8922	0	0.8176	0.9243
S	0	0.9470	0.9814	0	0.9462	0.9923

calculation results are for after their removal. Figure 8.2 portrays this latter case with before and after removal scatterplots; this is the worst case scenario for the five Polish employment categories. Except for these few prominent anomalies, the correspondences are quite good: in most cases, the bivariate slope coefficients are approximately 1, as are the R^2 values. The number of eigenvectors used to construct an eigenvector spatial filter (ESF) describing the SSRE terms range from 22 to 29 (from a candidate set of 85); the percentage of variance accounted for ranges

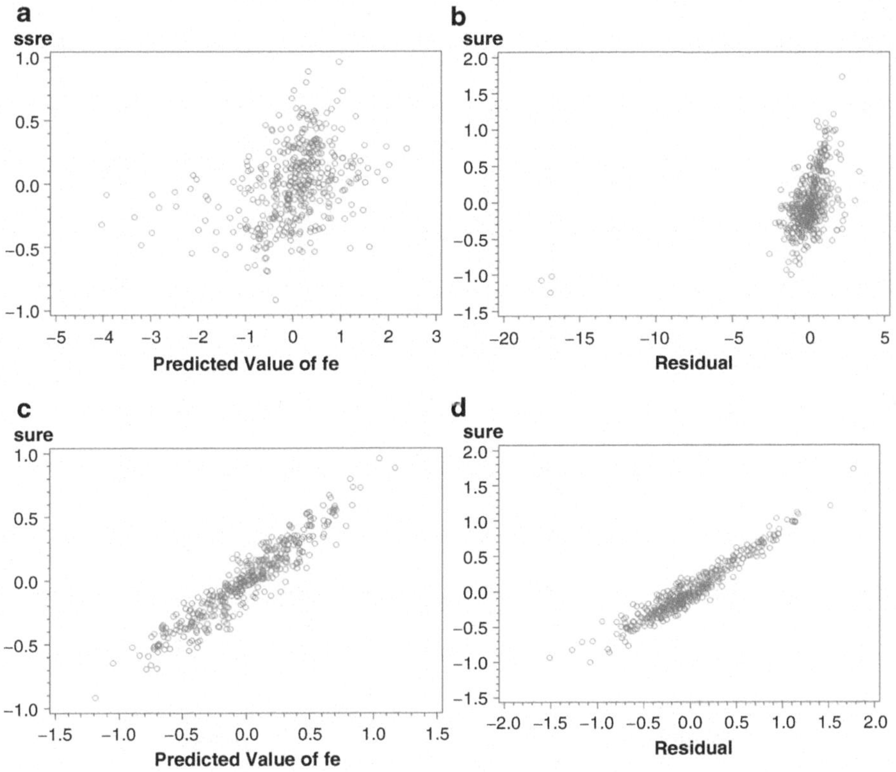

Fig. 8.2 $LQ_{finance}$ scatterplots. Top left (**a**): SSRE and SSFE. Top right (**b**): SURE and SUFE. Bottom left (**c**): SSRE and SSFE, three outliers removed. Bottom right (**d**): SURE and SUFE, three outliers removed

from 29% to 53%. The two anomalous FEs do not seem to relate to these summary features. As an aside, the intercepts are zero by construction.

Of note is that the FE anomalies are dramatically less than the remaining values in their respective sets. In addition, they correspond to the smallest values in their associated RE sets. The RE specification appears to shrink its estimates toward its mean of zero; the FE does not appear to either do or effectively do this shrinkage.

8.5 Estimating a SUFE Term: Selected Sensitivity Analyses

Chapter 4 presents methods of estimating RE terms, including when little or no repeated measures data are available. A SSRE term can be estimated by constructing an ESF based on regression model residuals. Furthermore, in the absence of repeated measures, one method exploits additional information furnished by prior distributions in a Bayesian analysis.

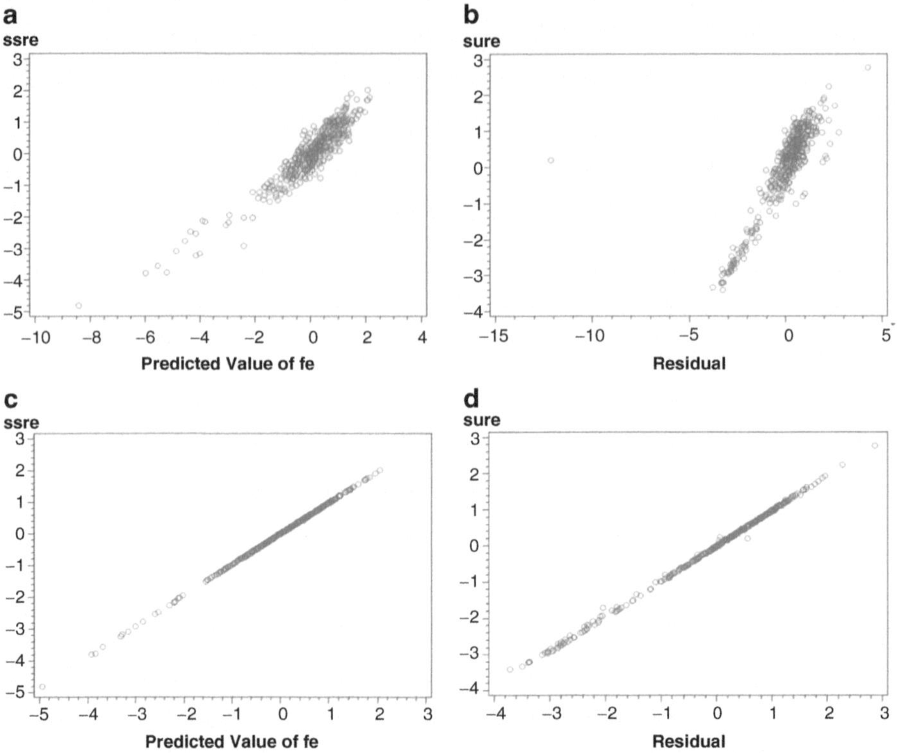

Fig. 8.3 $LQ_{agriculture}$ scatterplots. Top left (**a**): SSRE and SSFE. Top right (**b**): SURE and SUFE. Bottom left (**c**): SSRE and SSFE with the single outlier replaced by its imputed value. Bottom right (**d**): SURE and SUFE with the single outlier replaced by its imputed value

The preceding section presents another method for estimating these terms. But it appears to be plagued by potential extreme outliers. And it definitely requires repeated measures in order to compute the individual areal unit means. One way to handle extreme outliers is to exploit the normality assumption affiliated with RE. This can be achieved by replacing the extreme outliers with imputed values based upon fitting the well-behaved subset of estimated FE values to a normal frequency distribution by matching FE rank ordered values to their equivalent normal quantile values. Figure 8.3 portrays results of substituting the imputed value, −4.22, for the FE value, −20.58; the associated RE value is −4.60. Improvement of the relationship between the FE and RE values is dramatic.

Figure 8.2a, b confirm that $LQ_{finance}$ also is problematic. Three areal units have an RE value of −1.33 and an FE value of −20.88. Imputation calculations render the set of values {−1.04, −1.14, −1.33}; this set results in six (permutations) possible solutions.[1] Figure 8.4 portrays the scatterplots for the best solution. A

[1] All solutions were similar. The best solution has the following assignments: areal unit #17: −1.04, areal unit #70: −1.14, and areal unit #107: −1.33. Here, the criterion for "best" is the minimum sum of the absolute differences between the bivariate regression slope coefficients and 1.

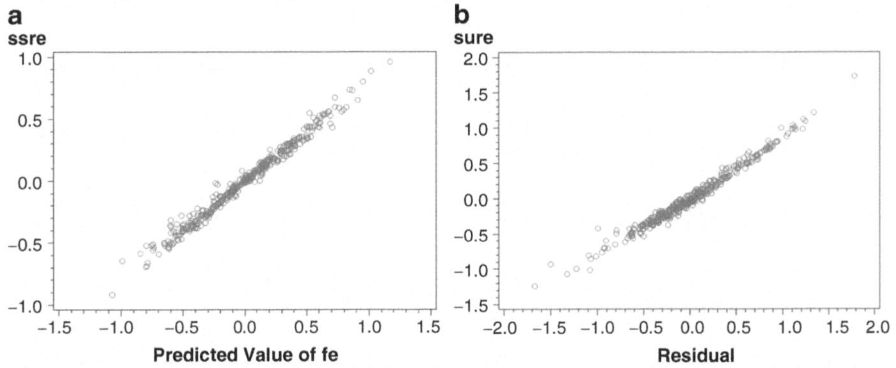

Fig. 8.4 $LQ_{finance}$ scatterplots. Left (**a**): SSRE and SSFE with the three outliers replaced by their imputed values. Right (**b**): SURE and SUFE with the three outliers replaced by their imputed values

comparison of Fig. 8.4a, b with Fig. 8.2c, d confirm that the imputations dramatically improve the relationship between the FE and RE values.

In addition, the SSFE and SUFE are relatively insensitive to the number of time periods used for repeated measures information (Fig. 8.5). $LQ_{industry}$ is the best case scenario for the categorized Polish employment data. Relationships between two and seven time period calculations are strong (Fig. 8.5a–c). $LQ_{finance}$ is the worst case scenario for the categorized Polish employment data. Again, relationships between two and seven time period calculations are strong (Fig. 8.5d–f).[2] In conclusion, the FE term seems rather insensitive to the number of time periods furnishing repeated measures information, at least for the categorized Polish employment data.

8.6 An Exploration of Interaction Terms

Two-way ANOVA requires interaction terms to be evaluated before interpreting their affiliated individual factor terms. Stepwise regression furnishes a technique for implementing such an evaluation when the number of variables equals or exceeds the number of observations. Here, the number of cross-classification indicator variables is $(T-1)(n-1) = 2274$. Using a Bonferroni adjustment to a 10% level of significance, and a beta-binomial probability model to account for overdispersion, the number of significant interaction indicator variables is 15 (i.e., less than 0.7%). This finding furnishes evidence that the semi-saturated FE specification is suitable, and that the maps appearing in Fig. 8.1c, d are meaningful.

[2]Now five, rather than three, extreme outliers exist in the data. These outliers have been removed when constructing Fig. 8.5d–f. Their estimation could result in 60 (permutations) possible solutions, whose best solution is not identified here.

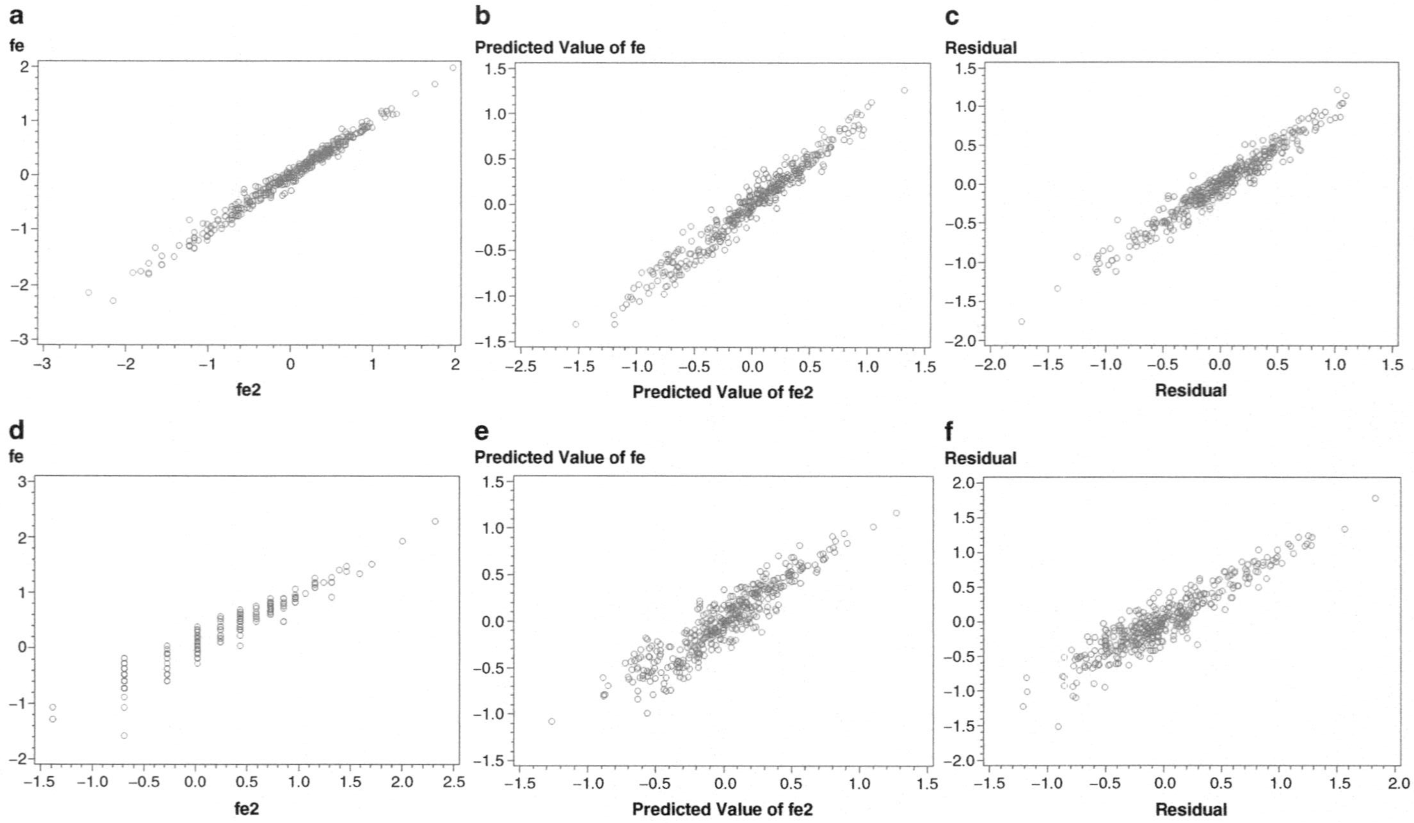

Fig. 8.5 Results for two and seven time periods. Top right (**a**); $LQ_{industry}$ FE. Top middle (**b**): $LQ_{industry}$ SSFE. Top right (**c**): $LQ_{industry}$ SUFE. Bottom right (**d**); $LQ_{finance}$ FE. Bottom middle (**e**): $LQ_{finance}$ SSFE. Bottom right (**f**): $LQ_{finance}$ SUFE

8.7 Concluding Comments

FE and RE specifications can comprise very similar value pairs, regardless of the magnitude of T. FE removes the areal unit time averages from a response variable and covariates at each point in time; RE removes only a fraction of these time averages. As this fraction goes to zero, paired FE and RE values converge (Wooldridge 2002, pp. 286–287). This convergence can occur as T goes to infinity, which is not the case for the categorized Polish employment data, or it can occur when within areal unit time series variance becomes increasingly large.

This chapter presents a clarification of FE, and a decomposition of FE into SSFE and SUFE, paralleling a relatively longstanding practice for RE; this is a somewhat novel contribution of this chapter. In addition, this chapter presents an assessment of what can be called semi-saturated FE specifications, which is an extension of Chap. 3 content. It explicitly outlines an evaluation tool for interaction effects in space–time data. Finally, this chapter presents additional analysis of the space–time categorized Polish employment data LQs, summarizing findings consistent with those appearing in, for example, Chap. 4.

References

Bell, A., & Jones, K. (2015). Explaining fixed effects: Random effects modeling of time-series cross-sectional and panel data. *Political Science Research and Methods, 3*, 133–153.

Clarke, P., Crawford, C., Steele, F., & Vignoles, A. (2010). *The choice between fixed and random effects models: Some considerations for educational research*. Discussion Paper No. 5287. Bonn: The Institute for the Study of Labor (IZA).

Nakamura, A., & Nakamura, M. (1981). On the relationships among several specification error tests presented by Durbin, Wu, and Hausman. *Econometrica, 49*, 1583–1588.

Wooldridge, J. (2002). *Econometric analysis of cross sectional and panel data*. Cambridge, MA: MIT Press.

Chapter 9
Spatial Autocorrelation and Spatial Interaction Gravity Models

9.1 Introduction

In the early 1970s, Curry (1972) argued that spatial autocorrelation (SA; i.e., local distance and configuration effects) and global distance decay (i.e., global distance effects) intermingle in the estimation of spatial interaction model specifications. Griffith and Jones (1980) confirmed this contention by analyzing journey-to-work flows in Canadian cities. But computing power and resources at that time did not support the type of model specification and estimation necessary to account for SA in spatial interaction data (Griffith 2007). This computer resources constraint no longer holds. During the last decade, SA components contained in journey-to-work flows have been addressed explicitly for a range of geographic landscapes and scales, including: Texas (Griffith 2009b); Pennsylvania (Griffith 2011); Germany (Griffith 2009a; Griffith and Chun 2015); and Washington, DC, Puerto Rico, Chicago, and Dallas-Fort Worth (Griffith and Chun 2015). This chapter outlines a methodology used to account for SA in gravity model specifications and adds a Polish journey-to-work example to the preceding collection of empirical case studies.

9.2 The Doubly Constrained Gravity Model: A Poisson Specification That Accounts for Spatial Autocorrelation

A doubly constrained gravity model includes an indicator variable for each origin and each destination areal unit, which ensures that the observed and predicted number of workers leaving an origin, and the observed and predicted number of works arriving at a destination, are exactly the same (Griffith and Fischer 2013).

D. A. Griffith, J. H. P. Paelinck, *Morphisms for Quantitative Spatial Analysis*,
Advanced Studies in Theoretical and Applied Econometrics 51,
https://doi.org/10.1007/978-3-319-72553-6_9

The coefficients of these indicator variables are the log-balancing factors in the Wilson entropy-based model specification. This SA-adjusted specification may be written as

$$F_{ij} = O_i^{\delta} D_j^{\lambda} \exp\left(\alpha + \sum_{h=1}^{n} I_{i,h,\mathrm{o}} \beta_{h,\mathrm{o}} + \sum_{k=1}^{n} I_{j,k,\mathrm{d}} \beta_{k,\mathrm{d}} - \gamma d_{ij} + \beta I_{d_{ij}=0} + \rho_{ij} \right), \quad (9.1)$$

where

F_{ij} is the journey-to-work flow between origin i and destination j,
O_i and D_j are the total number, respectively, of workers in origin i and of jobs in destination j that constitute flows within the system and enters a Poisson regression model specification as an offset variable LN(O_i)+LN(D_j),
exp($\beta_{h,\ \mathrm{o}}$) and exp($\beta_{k,\ \mathrm{d}}$), respectively, are the origin h (conventionally denoted by A_i) and the destination k (conventionally denoted by B_j) balancing factors, with the nth areal unit assigned $\beta_{n,\ \mathrm{o}}$ (i.e., $A_i = 1$) and $\beta_{n,\ \mathrm{d}}$ (i.e., $B_j = 1$) values of 0 (to avoid perfect multicollinearity in the specification)—the sum of these two values is absorbed into the intercept term,
$I_{i,h,\mathrm{o}}$ and $I_{j,k,\mathrm{d}}$ denote the binary 0-1 indicator variables for, respectively, the origin and destination areal units associated with areal unit i [i.e., indicator variables with n values of 1, and $n(n-1)$ values of 0, such that $I_{i,h,\mathrm{o}} = 1$ when $i = h$, and $I_{j,k,\mathrm{d}} = 1$ when $j = k$],
d_{ij} is the distance separating origin i and destination j (measured here between areal unit centroids),
β is the distance decay differential for within areal unit flows,
ρ_{ij} is the SA term representing dependencies between flows from nearby origin areal units of i to nearby destination areal units of j, and
exp(α) and γ, respectively, are the constant of proportionality and the inter-areal unit distance decay parameter; δ and λ are traditional gravity model parameters.

The sets of balancing factors exp($\beta_{h,\ \mathrm{o}}$) and exp($\beta_{k,\ \mathrm{d}}$), respectively, ensure that $\sum_{i=1}^{n} \widehat{F}_{ij} = O_i$ and that $\sum_{j=1}^{n} \widehat{F}_{ij} = D_j$, where $\widehat{F}_{ij}$ denotes the predicted flow between origin i and destination j. These values appear only when $I_{i,h,\mathrm{o}}$ and $I_{j,k,\mathrm{d}}$ equal 1. Otherwise, when $I_{i,h,\mathrm{o}}$ and $I_{j,k,\mathrm{d}}$ equal 0, the factors become exp(0) = 1, which disappear in the specification because any value multiplied by 1 equals itself. Meanwhile, Curry (1972) argues that inclusion of the SA term ρ_{ij} results in a substantial change in the estimate of γ. In other words, the absence of this term in a specification results in a convolution of local and global distance effects.

9.3 Modeling Spatial Autocorrelation

In this chapter, accounting for SA in spatial interaction flows is achieved with an eigenvector spatial filter (ESF) specification. ESF methodology furnishes an approach that can account for SA without having the flows counts variable Y on both sides of an equal sign (like in an autoregressive specification; see LeSage and Pace 2008). It also can be used to analyze flows data described by a Poisson probability model specification, allowing components of SA to be included in a model specification. This latter result is achieved by constructing a linear combination of a subset of the eigenvectors of an n-by-n modified geographic weights matrix $\mathbf{C}_n$—an ESF. This subset can be determined with stepwise Poisson regression based upon statistical significance.

The modified geographic weights matrix for a given surface partitioning from which eigenfunctions are extracted is $(\mathbf{I} - \mathbf{11}'/n)\mathbf{C}_n(\mathbf{I} - \mathbf{11}'/n)$—where $\mathbf{I}$ is an n-by-n identity matrix, $\mathbf{1}$ is an n-by-1 vector of 1s, and $'$ denotes the matrix transpose operation. This matrix expression appears in the numerator of the Moran Coefficient (MC) SA index. The eigenvectors of this matrix exhibit the following property when mapped:

The first eigenvector, say $\mathbf{E}_1$, is the set of real numbers that has the largest MC value achievable by any set of real numbers for the spatial arrangement defined by the geographic connectivity matrix $\mathbf{C}_n$. The second eigenvector is the set of real numbers that has the largest achievable MC value by any set that is uncorrelated with $\mathbf{E}_1$. The third eigenvector is the third such set of values. And so on through $\mathbf{E}_n$, the set of real numbers that has the largest negative MC value achievable by any set that is uncorrelated with the preceding $(n-1)$ eigenvectors.

As such, these eigenvectors furnish distinct map pattern descriptions of latent SA in georeferenced variables because they are both orthogonal and uncorrelated. Their corresponding eigenvalues index the nature and degree of SA portrayed by each eigenvector (Tiefelsdorf and Boots 1995), which can be standardized by dividing by the largest MC value (which often is greater than 1 and occasionally is less than 1). As an aside, this decomposition could be based upon the Geary Ratio (GR) SA index.

The SA of interest here is contained in the network of flows that constitutes spatial interaction—the flows themselves are spatially dependent. Specifically, flows between a given pair of areal units are correlated with flows between areal units that are nearby the origin and the destination units in the pair. These origin/destination paired variables can be linked to the individual flows data dyads through Kronecker products, which results in an n^2-by-n^2 connectivity matrix for flows that is asymptotically correct: $\mathbf{C}_{n^2} = \mathbf{C}_n \otimes \mathbf{C}_n$, where $\mathbf{C}_{n^2}$ is the n^2-by-n^2binary 0-1 connectivity matrix for the n^2 spatial interaction flows, and $\mathbf{C}_n$ is the n-by-n connectivity matrix for the tessellation of n origins/destinations. As with most geographic weights matrices, $\mathbf{C}_{n^2}$ is very sparse. It also is quite sizeable. This Kronecker product specification yields the origin and destination variates $\mathbf{E}_i \otimes \mathbf{E}_j$ and $\mathbf{E}_j \otimes \mathbf{E}_i$ for constructing spatial filters. For example, the corresponding eigenvector value for origin i in a given vector is multiplied by each of the values of a given eigenvector j, one value for each of the n destinations. This matrix operation yields an n^2-by-1 vector (eigenvector j is concatenated n times, with each one of the set of n values being multiplied, in turn, by a value in eigenvector i).

These Kronecker products are similar to interaction terms in classical regression. Because a minimum level of SA should be represented by a given eigenvector, say

0.25 (i.e., roughly 5% of the variance of a given eigenvector's values is attributable to redundant information resulting from the presence of SA), and the eigenvalues of $\mathbf{C}_n \otimes \mathbf{C}_n$ are the pairwise products of the individual eigenvalues of $\mathbf{C}_n$, this threshold value can be increased to 0.5 (i.e., $0.5^2 = 0.25$) for each of the eigenvectors in a Kronecker product when the number of eigenvectors is large.

9.4 Spatial Autocorrelation and Provincial-Level Journey-to-Work Flows

In 2011, Poland had 2,758,099 workers traveling between their residences and their jobs.[1] Poland is divided into 16 provinces (i.e., viovodeships), which are very large (i.e., a coarse geographic resolution; see Fig. 9.1a) areal units. Inter-provincial distances between provincial areal unit centroids (Fig. 9.1b), supplemented with intra-provincial distances measured as $\frac{2}{3}\sqrt{\text{Area}/\pi}$ quantify separation here. The total number of flows is $16^2 = 256$, of which 27 are zero.

Five different gravity model specifications were estimated to describe these data (Table 9.1). Each is a particular version of Eq. (9.1). The inter-areal unit global distance coefficient is negative, with adjustment for SA (via 8 Kronecker product eigenvectors from a candidate set of 16 eigenvectors, which is a subset of 256 vectors) reducing it to nearly half of its traditional gravity model estimate. The intra-

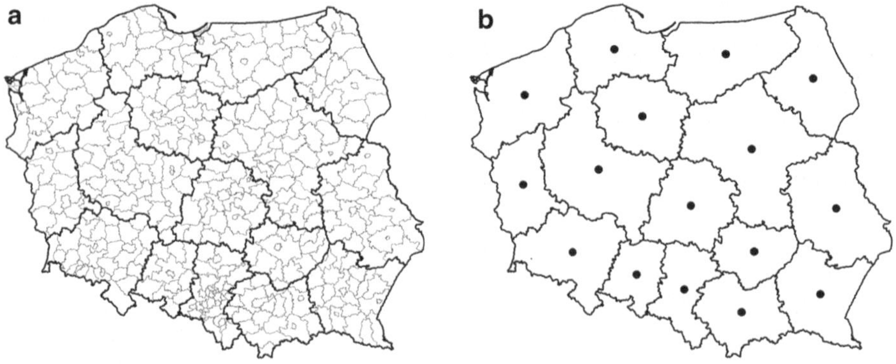

Fig. 9.1 The spatial structure of Poland. Left (**a**) provinces boundaries (heavy black lines) and county boundaries (gray lines). Right (**b**): provincial geometric centroics (black filled circles)

[1]These commuting data were compiled based on administrative sources. Commuters are salaried employees (staff), whose place of employment is outside the administrative borders of their municipality of residence. These data include only flows of more than nine people. See http://stat.gov.pl/spisy-powszechne/nsp-2011/nsp-2011-wyniki/dojazdy-do-pracy-w-polsce-wyniki-nsp-2011,9,1.html.

Table 9.1 Provincial-resolution spatial interaction gravity model estimation summary statistics

	Distance		Bivariate regression			Exponent		Over-dispersion (Pearson)
Specification	Inter-	Intra-	a	b	Pseudo-R^2	Origin weight	Destination weight	
Traditional	0.0226	***	−256	1.0247	0.9839	0.3003	0.7515	116
Conventional doubly constrained	0.0246	***	−318	1.0296	0.9918	1	1	138
Constant intra-distance	0.0079	3.1728	117	0.9892	0.9984	1	1	30
Variable intra-distance	0.0088	0.0060	49	0.9954	0.9976	1	1	49
Accounting for SA	0.0130	0.0050	46	0.9957	0.9984	1	1	33

*** denotes "does not exist"

areal unit global distance coefficient is positive, indicating that as province size increases, the number of within-province journey-to-work flows tends to increase; this is a size effect. The constant intra-distance version includes an indicator variable differentiating within and between province flows; the variable intra-distance specification multiplies this indicator variable by $\frac{2}{3}\sqrt{\text{Area}/\pi}$. The observed-versus-predicted flows goodness-of-fit coupled with relatively low over-dispersion indicates that the full model specification given by Eq. (9.1) furnishes the best Polish journey-to-work data description.

Figure 9.2 presents scatterplots for the two extreme cases of these specifications. The roughly dozen substantially larger flows clarify why the pseudo-R^2 values in Table 9.1 are so close to 1. Refining the model specification to account for SA better aligns the smaller observed and predicted flows, as well as one of the three largest flows. These findings are consistent with and corroborate those already appearing in the literature for other geographic landscapes (see Sect. 9.1).

Figure 9.3 portrays features of the SA components of Poland's journey-to-work flows. Figure 9.3a, b are the geographic distributions of the spatially structured (SSAI) and spatially unstructured (SUAI) parts of the concatenated origin indicator variable coefficients (augmented with a 0 for the 16th province). The SSAI part is equivalent to the eigenvector representing maximum negative SA (MC = −0.54; GR = 1.65). This single eigenvector accounts for nearly 50% of the origin weights' geographic variability. The geographic distribution of the propensity to supply workers exhibits moderate negative SA[2]; it displays latent aspects of spatial competition. Figure 9.3c, d are the geographic distributions of the spatially structured (SSBJ) and spatially unstructured (SUBJ) parts of the concatenated destination indicator variable coefficients (augmented with a 0 for the 16th province). The

[2]Spatial scientists rarely encounter negative SA. For these origin weights, MC = −0.22 and GR = 1.27.

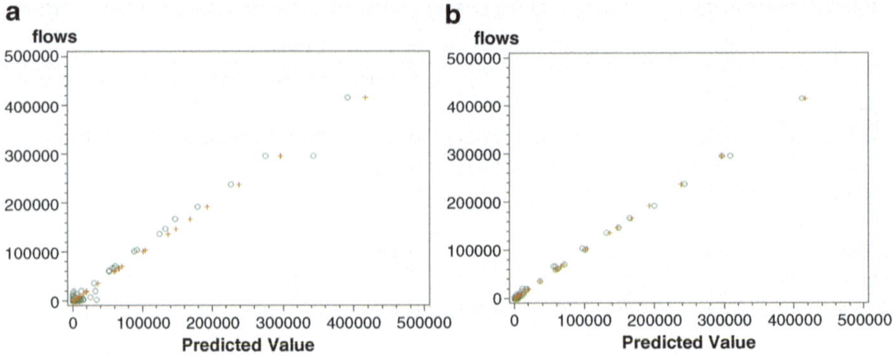

Fig. 9.2 Scatterplots of observed versus predicted flows (open circles); pluses denote a perfect fit line. Left (**a**): the conventional specification. Right (**b**): the SA-adjusted specification

SSBJ part is equivalent to the eigenvector representing near-maximum positive SA (MC = 0.61; GR = 0.34). This single eigenvector accounts for nearly 40% of the destination weights' geographic variability. The geographic distribution of the propensity to supply jobs exhibits weak positive SA[3]; it displays latent aspects of spatial agglomeration effects. Figure 9.3e portrays the origin network flows SA component, which contains weak positive SA (MC = 0.06, GR = 0.84). The magnitude of the outflow of workers from a province tends to be only very slightly similar to the outflows from its neighboring provinces; this weak tendency may well result, in part, from competitive suppression related to negative SA contained in the geographic distribution of the origin propensity weights. Figure 9.3f portrays the destination network flows SA component, which contains moderate[4] positive SA (MC = 0.33, GR = 0.65). The magnitude of the inflow of workers to a province tends to be similar to the inflows to its neighboring provinces; this moderate tendency may well result, in part, from agglomeration multiplier effects related to positive SA contained in the geographic distribution of the destination propensity weights.

As an aside, Griffith and Fischer (2013) link the spatially structured and unstructured components discussed here to the SSRE and SURE topics addressed in Chap. 4.

9.5 Infill and Increasing Domain Analyses

A comparative analysis of provincial, county, and commune level (Figs. 9.1a and 9.4b) journey-to-work flows involves a shift in geographic resolution (the infill perspective). Increasing geographic scale from a single province, say Łódź

[3]For these destination weights, MC = 0.12 and GR = 0.93.

[4]For the Polish province surface partitioning, the maximum positive SA possible is MC = 0.76 and GR = 0.27.

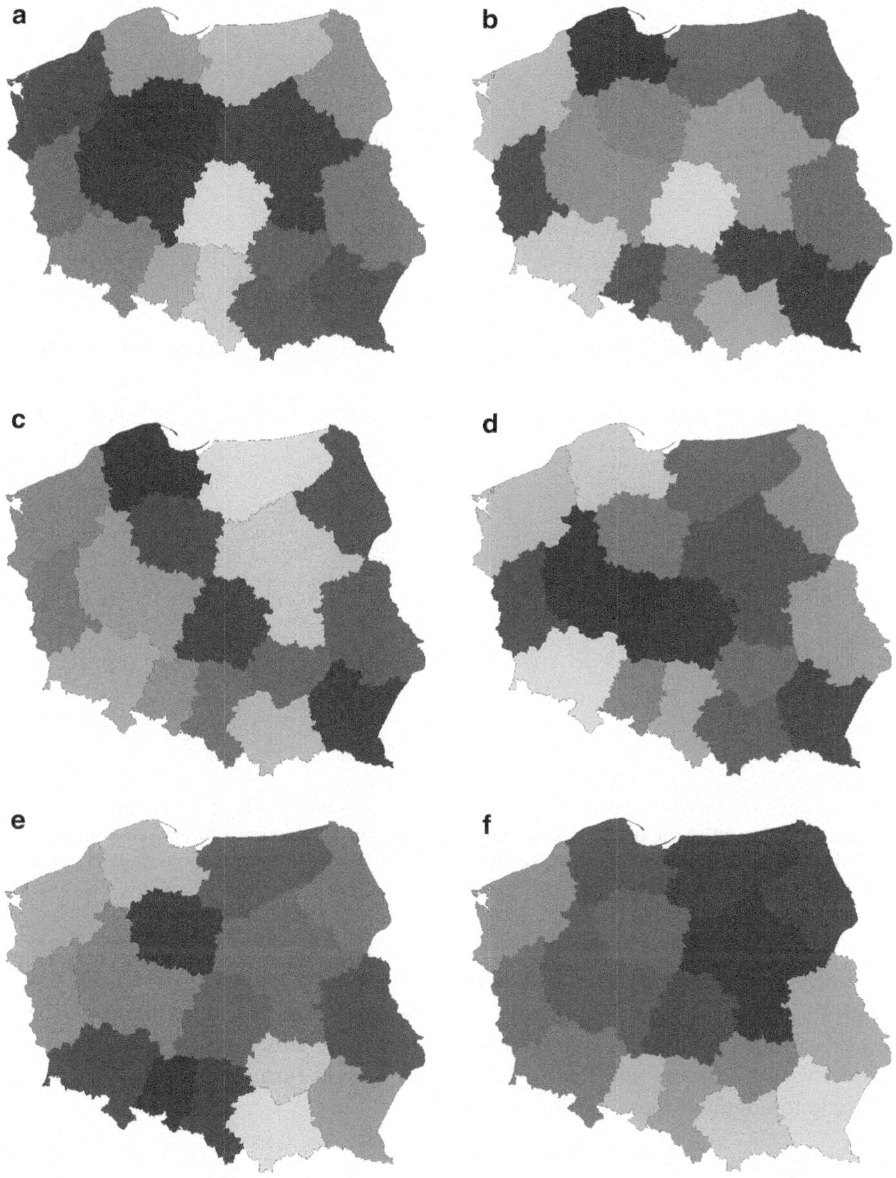

Fig. 9.3 Province resolution SA components; grayscale is directly proportional to magnitude of quantities. Top left (**a**): SSAI. Top right (**b**): SUAI. Middle left (**c**): SSBJ. Middle right (**d**): SUBJ. Bottom left (**e**): median origin network SA ESF. Bottom right (**f**): median destination network SA ESF

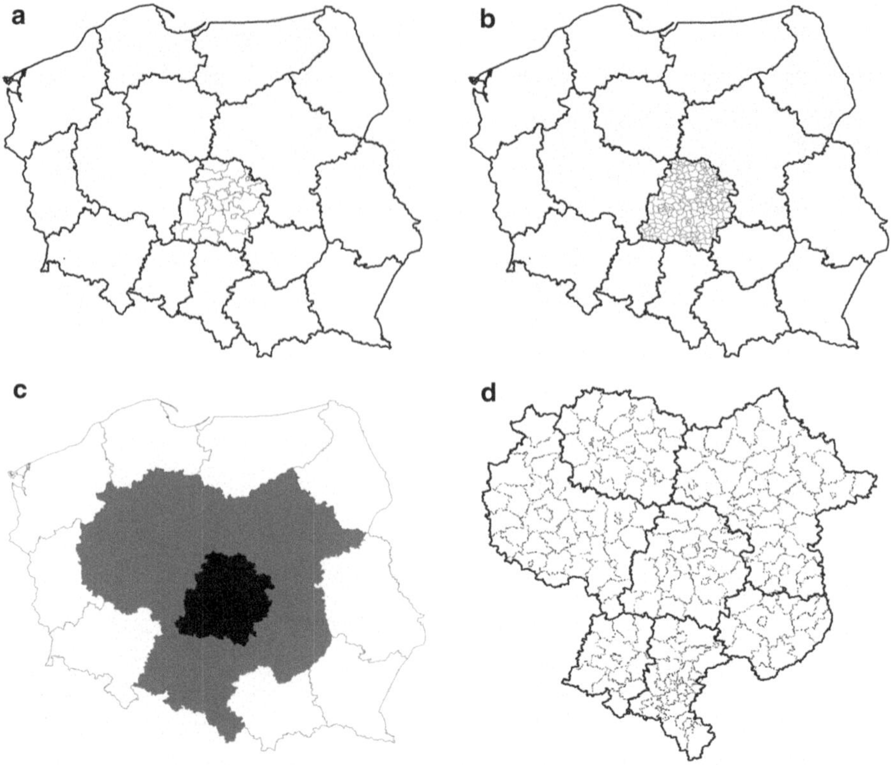

Fig. 9.4 Infill and increasing domain geographic landscapes based upon the Łódź Province. Top left (**a**): county partitioning of Łódź. Top right (**b**): commune partitioning of Łódź. Bottom left (**c**): Łódź (black), its six surrounding provinces (gray), and the rest of Poland. Bottom right (**d**): county partitioning of Łódź and its six surrounding provinces

(centrally located; Fig. 9.4c), to it combined with its six adjacent provinces, to all 16 provinces, involves a shift in geographic scale (the increasing domain perspective). Griffith and Chun (2015) address both of these situations in the context of, respectively, Germany and Puerto Rico. This section furnishes another empirical example to augment these earlier findings.

9.5.1 A Comparative Infill Analysis of Journey-to-Work Flows

The Łódź Province is centrally located. It contains 24 counties (Fig. 9.4a) and 176 communes (Fig. 9.4b). The province has 136,635 internal journey-to-work flows, leaks 30,968 workers to jobs outside of the province, and attracts 22,969

Table 9.2 County-resolution spatial interaction gravity model estimation summary statistics for Łódź Province

	Distance		Bivariate regression			Exponents		Over-dispersion (Pearson)
Specification	Inter-	Intra-	*a*	*b*	Pseudo-R^2	Origin weight	Destination weight	
Traditional	0.0791	***	21	0.9130	0.8493	0.1669	1.0422	13
Conventional doubly constrained	0.0954	***	3	0.9870	0.9754	1	1	8
Constant intra-distance	0.0858	0.5505	2	0.9909	0.9773	1	1	7
Variable intra-distance	0.0885	0.0355	3	0.9889	0.9757	1	1	8
Accounting for SA	0.0982	−0.0111	−0	1.0016	0.9900	1	1	7

*** denotes "does not exist"

workers to jobs within the province. Of the 576 county spatial interaction dyads, 377 are zero.

Table 9.2 summarizes results comparable to Table 9.1.[5] Again, the full Eq. (9.1) specification furnishes the best description of these data. The network flows SA component is an ESF constructed with 13 Kronecker product eigenvectors (from a candidate set of 36 eigenvectors based on individual eigenvector MC values of at least 0.25). One difference between Tables 9.1 and 9.2 is the negative sign for the intra-distance variable, but this might be an artifact arising from the incompleteness of the Polish spatial interaction data. Without adjusting for SA, global distance decay is over-estimated for the provincial level analysis, but under-estimated for the given county level analysis. As an aside, the size effect estimates (i.e., origin and destination weight exponents) for the traditional model specification probably are poor, given that a city with 100,000s, and two cities with 10,000s, of people are not included in the analysis (because of their missing journey-to-work flows).

Equation (9.1) supports maximum likelihood (ML) estimation (i.e., conditional mean imputations) of intra-flows, which are as follows (for 2011):

County	Out-flow	ML estimate	Total	Total employment
Łódź	18,056	218,076	236,132	228,805
Piotrków Trybunalski	4134	5766	9900	24,901
Skierniewice	3858	2762	6620	12,707

[5]Because the Polish census defines a commuter to be a salaried employee (staff member) whose place of employment is outside the administrative borders of his/her municipality of residence, three counties, which actually are cities—Skierniewice, Łódź, and Piotrków Trybunalski—have artificial zero flows. To preserve comparability, these three flows were differentiated from the remaining flows by an indicator variable.

Table 9.3 Inter-areal unit spatial interaction gravity model estimation summary statistics for Łódź Province

Specification	Inter-areal unit distance	Bivariate regression *a*	*b*	Pseudo-R^2	Exponent Origin weight	Destination weight	Over-dispersion (Pearson)
County-resolution							
Traditional	0.0583	13	0.9089	0.8333	0.4233	0.8497	11
Conventional doubly constrained	0.0881	2	0.9824	0.9754	1	1	7
Accounting for SA	0.0915	1	0.9923	0.9848	1	1	6
Commune-resolution							
Traditional	0.0854	0	0.8534	0.7327	0.7920	0.9330	3
Conventional doubly constrained	0.1243	−0	1.0486	0.9150	1	1	4
Accounting for SA	0.1364	0	0.9972	0.9898	1	1	1

These conditional expectations yield total quantities that compare favorably with their respective recorded total employment counts.

Next, an analysis of only the inter-county flows allows a comparison between the county- and the commune-resolution journey-to-work flows. This restriction does not change the total number of journeys-to-work, but reduces the number of county-resolution dyads to 552, of which 374 have a zero flow. County-resolution results reported in Table 9.3 are similar to their counterparts appearing in Table 9.2; here the full Eq. (9.1) results involve 10 Kronecker product eigenvectors (from a candidate set of 36 eigenvectors based on individual eigenvector MC values of at least 0.25).

The commune-resolution data are sufficiently sparse that they introduce several challenges. These data contain 30,800 inter-commune dyads, 29,990 of which have a zero flow; in other words, only 810 of the dyads have nonzero flows. Furthermore, 11,504 of the dyads have either its origin or its destination commune with zero workers/jobs; in other words, Eq. (9.1) yields a predicted value of zero (i.e., the product term includes one or two zeroes) for these zero flows. But these data cannot be included in the Poisson estimation because they insert the logarithm of zero into the offset variable. Here, the full Eq. (9.1) results involve 258 Kronecker product eigenvectors (from a candidate set of 529 eigenvectors based on individual eigenvector MC values of at least 0.50). As with the provincial- and county-resolution cases, the full Eq. (9.1) outperforms the SA-unadjusted specifications. For these fine geographic resolution data, adjusting for SA removes over-dispersion and furnishes a goodness-of-fit line whose intercept and slope are nearly identical to their respective theoretical counterparts of zero and one. One implication here is that geographic aggregation can create excess Poisson variation.

Findings summarized in this section render two principal implications. First, Eq. (9.1) furnishes an extremely good description of Polish journey-to-work data. Second, Tables 9.1 and 9.3 confirm that the global distance decay exponent increases with geographic resolution, and that its estimation ignoring SA effects yields a biased value: the coarsest geographic resolution produces an inflated, whereas the finer geographic resolution produces a deflated, global distance decay estimate.

9.5.2 A Comparative Increasing Domain Analysis of Journey-to-Work Flows

As noted in the preceding section, the Łódź Province is centrally located and contains 24 counties (Fig. 9.4a). Łódź together with its six adjacent provinces contains 186 counties (Fig. 9.4d). Poland contains 380 counties (Fig. 9.1a).[6] As a geographic domain increases, the percentage of zeroes in its flows matrix tends to increase: 374 of 552 for the Łódź Province (67.8%), 31,483 of 34,410 for Łódź together with its six adjacent provinces (91.5%), and 137,265 of 144,020 for Poland (95.3%). One outcome is that over-dispersion tends to increase. At least for this dataset, neither a negative binomial nor a zero-inflated Poisson specification successfully accounts for this excess Poisson variation (Table 9.4). As an aside, 1,830,580 of the 2,758,099 commuters work in a county different from where they reside (i.e., inter-county journey-to-work flows).

Figure 9.5 portrays the scatterplot of observed versus predicted flows (similar to Fig. 9.2) for the entire set of 380 counties of Poland.

As the geographic domain increases, while the geographic resolution remains the same, the SA-adjusted global distance decay parameter estimate decreases (from 0.0915 to 0.0389, then to 0.0313).

Finally, in contrast with Fig. 9.3, Fig. 9.6 portrays the geographic distributions of the various SA components for the county geographic resolution. The origin log-balancing factors (Fig. 9.6a, b) have a SA component containing moderate-to-strong positive SA (MC = 0.88); SA accounts for roughly 42% of the geographic variability in these values. The destination log-balancing factors (Fig. 9.6c, d) also have a SA component containing moderate-to-strong positive SA (MC = 0.94); SA accounts for roughly 36% of the geographic variability in these values. The origin and destination network SA components contain moderate positive SA (respectively, MC = 0.67 and MC = 0.78).

[6]As with the employment data (see Chap. 1), the journey-to-work data for Wałbrzych are zeroes and are wrong. Because a time series of journey-to-work data does not exist, these flows data were not corrected with imputations. Rather, this footnote acknowledges this data error; the flows data remained linked to Wałbrzyski for this spatial interaction analysis.

Table 9.4 Inter-county spatial interaction gravity model estimation summary statistics for an increasing provincial domain from Łódź Province

Specification	Inter-areal unit distance	Bivariate regression a	b	Pseudo-R^2	Exponent Origin weight	Destination weight	Over-dispersion (Pearson)
Łódź and its six adjacent provinces							
Traditional	0.0325	3	0.8946	0.6615	0.5757	1.1422	31
Conventional doubly constrained	0.0439	0	0.9915	0.9010	1	1	582
Accounting for SA	0.0389	0	0.9971	0.9588	1	1	234
All 16 Polish provinces							
Traditional	0.0298	2	0.8623	0.5423	0.4890	1.1465	151
Conventional doubly constrained	0.0359	−0	1.0325	0.8359	1	1	1043
Accounting for SA[a]	0.0313	0	0.9971	0.9050	1	1	503

[a]Given the data error problem, this estimation was not optimized. The SA component ρ_{ij} was defined with the 400 Kronecker product eigenvectors constructed using the 20 eigenvectors of matrix **C** with MC > 0.75

9.6 Concluding Comments

Both theoretical and prior empirical evidence contend that SA can bias the estimated global distance decay value governing spatial interaction. This chapter furnishes yet another example demonstrating the nature and degree of this bias. The Polish journey-to-work data corroborate this finding.

This chapter presents additional infill and increasing domain gravity model estimation results; this is a useful research replication contribution of this chapter. While geographic scale is held constant, as geographic resolution becomes

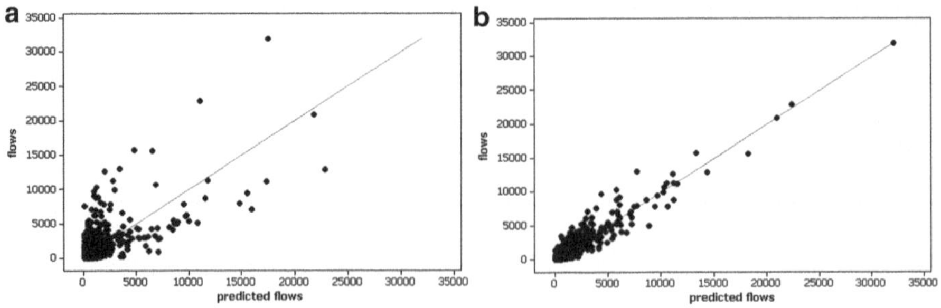

Fig. 9.5 Scatterplots of observed (vertical axis) versus predicted (horizontal axis). Left (**a**): traditional gravity model results. Right (**b**): SA-adjusted gravity model results

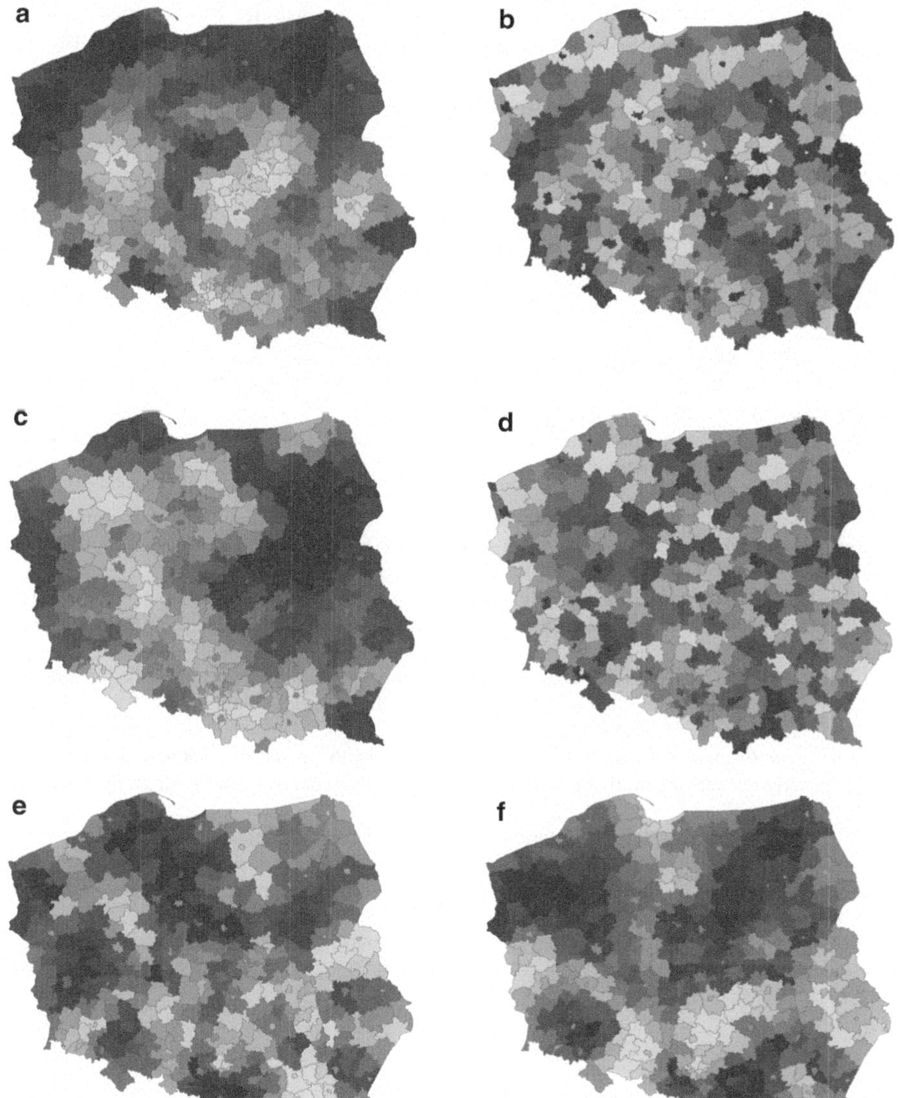

Fig. 9.6 County resolution SA components; grayscale is directly proportional to magnitude of quantities. Top left (**a**): SSAI. Top right (**b**): SUAI. Middle left (**c**): SSBJ. Middle right (**d**): SUBJ. Bottom left (**e**): median origin network SA ESF. Bottom right (**f**): median destination network SA ESF

increasingly finer, the global distance decay parameter tends to increase. While geographic resolution is held constant, as geographic scale increases, the global distance decay parameter tends to decrease. In addition, this chapter presents a puzzling finding that is a novel contribution: when a gravity model specification accounts for SA, extra Poisson variation increases (at least as measured by the Pearson deviance statistic). This finding is counterintuitive and merits additional future research. One additional research topic meriting attention concerns the Polish journey-to-work data: the analysis summarized in Sect. 9.5 needs to be redone with corrected data for Wałbrzych and Wałbrzyski.

References

Curry, L. (1972). Spatial analysis of gravity flows. *Regional Studies, 6*, 131–147.

Griffith, D. (2007). Spatial structure and spatial interaction: 25 years later. *The Review of Regional Studies, 37*(1), 28–38.

Griffith, D. (2009a). Modeling spatial autocorrelation in spatial interaction data: Empirical evidence from 2002 Germany journey-to-work flows. *Journal of Geographical Systems, 11*, 117–140.

Griffith, D. (2009b). Spatial autocorrelation in spatial interaction: Complexity-to-simplicity in journey-to-work flows. In P. Nijkamp & A. Reggiani (Eds.), *Complexity and spatial networks: In search of simplicity* (pp. 221–237). Berlin: Springer.

Griffith, D. (2011). Visualizing analytical spatial autocorrelation components latent in spatial interaction data: An eigenvector spatial filter approach. *Computers, Environment and Urban Systems, 35*, 140–149.

Griffith, D., & Chun, Y. (2015). Spatial autocorrelation in spatial interactions models: Geographic scale and resolution implications for network resilience and vulnerability. *Networks and Spatial Economics, 15*, 337–365.

Griffith, D., & Fischer, M. (2013). Constrained variants of the gravity model and spatial dependence: Model specification and estimation issues. *Journal of Geographical Systems, 15*, 291–317.

Griffith, D., & Jones, K. (1980). Explorations into the relationship between spatial structure and spatial interaction. *Environment and Planning A, 12*, 187–201.

LeSage, J., & Pace, R. (2008). Spatial econometric modelling of origin-destination flows. *Journal of Regional Science, 48*, 941–967.

Tiefelsdorf, M., & Boots, B. (1995). The exact distribution of Moran's I. *Environment and Planning A, 27*, 985–999.

Chapter 10
General Conclusions About Spatial Statistics

10.1 Introduction

Griffith and Paelinck (2011) present an expanded version of their 2007 paper, with the first part of their book focusing on the following nonstandard spatial statistics topics: (1) individual versus ecological analyses, (2) statistical models for spatial data: some linkages and communalities, (3) frequency distributions for simulated spatially autocorrelated random variable, (4) understanding correlations among spatial random variables, (5) spatially structured random effects: a comparison of three popular specifications, (6) spatial filter versus conventional spatial model specifications: some comparisons, and (7) the role of spatial autocorrelation in prioritizing sites within a geographic landscape. Adopting a parallel structure in this book, its first part extends the previous spatial statistics treatments by emphasizing morphisms.

10.2 Spatial Autocorrelation and the *p*-Median Problem

To date, spatial statistics and spatial optimization research have evolved separately with relatively few interfaces. Statistics and operations research are considered natural companions in a number of academic settings (e.g., Carnegie Mellon University, Massachusetts Institute of Technology, University of Edinburgh, University of North Carolina/Chapel Hill), in part because each subject furnishes advanced analytical methods for organizing, modeling, solving, and analyzing a wide variety of data-driven practical problems. This companionship easily should extend to spatial statistics and spatial optimization, but thus far essentially has not. Chapter 2 establishes a foundation for at least initializing such a partnering.

D. A. Griffith, J. H. P. Paelinck, *Morphisms for Quantitative Spatial Analysis*, Advanced Studies in Theoretical and Applied Econometrics 51,
https://doi.org/10.1007/978-3-319-72553-6_10

One critical concept for an interface between these two disciplines is spatial autocorrelation, and a relevant optimization problem is location–allocation (LA). Kuenne and Soland (1971, p. 198) note the following about an exact solution for LA problems:

An important input to the algorithm, both in terms of the computational time required for convergence and the solution value obtained, is the set of initial source locations.

Not only does spatial autocorrelation support the imputation of any missing demand weights in a LA problem, but it also furnishes local statistics that can help determine an initial starting solution of source locations. This latter spatial cluster identification tool may prove to be particularly useful with heuristic LA algorithms such as ALTERN, which Brimberg and Hodgson (2011) note is widely used, even today. Hansen et al. (1998) emphasize that ALTERN has many weaknesses, but is a good heuristic in its multi-start version, which Brimberg and Hodgson (2011) describe as state-of-the-art.

Chapter 2 addresses the incapacitated LA problem. Future work needs to address the capacitated LA problems, as well as location covering problems, sequential location problems, hierarchical location problems, and stochastic location problems (Farahani and Hekmatfar 2009; Eiselt and Marianov 2011).

10.3 Space–Time Autocorrelation

Space–time data analyses are increasing in number because such data are becoming more plentiful and accessible, and hence more popular (Cressie and Wikle 2011). Chapter 3 begins by revisiting the literature of the early 1980s, focusing on the isomorphism between the Moran Coefficient (MC) and the MC_{ST} (i.e., space versus space–time) and then addresses the relatively new eigenvector space–time filtering methodology (Griffith 2012), with special reference to how it might help remedy omitted variable bias in space–time model specifications.

Space–time forecasting requires quite lengthy time series data to ensure high quality predictions. Such data tend to be characterized by the dominance of temporal dependence, emphasizing the importance of a sufficiently large time sequence. In practice, regional space–time data often have a relative short time series, a situation that will change with the passing of time, increasing chances that current regional forecasts might result in unreliable predictions. Griffith and Chun (2014) propose a method to improve regional forecasts by incorporating spatial autocorrelation in a generalized linear mixed model framework coupled with eigenvector spatial filtering. Chapter 3 adds to that methodology, grounding it in earlier work based upon a space–time MC, and contributing to solving some of the challenges outlined in Griffith (2010b). Specifically, because specifications of panel data models struggle to accommodate lag relationships in both space and time, the eigenvector spatial filter specification offers one way to achieve this end, allowing, as in Griffith and Chun (2014), spatial autocorrelation to be accounted for by a spatially structured random effects term and serial correlation to be accounted for by a time lag. Another challenge alluded to in Chap. 3 pertains to non-normal random variables and could benefit from the static case analyzed in Griffith (2010a). Yet a third challenge concerns the considerable heterogeneity, in terms of nonconstant variance for auto-normal data and overdispersion for auto-binomial and Poisson data, displayed by most space–time data. Once again, eigenvector space–time filtering furnishes a way to capture much of this heterogeneity. These particular challenges highlight the importance of the omitted variables evaluation presented in Chap. 3.

10.4 The Relative Importance of Spatial and Temporal Autocorrelation

Serial dependency tends to dominate autocorrelation in space–time data, with this dominance partly arising from spatial autocorrelation being two-directional and two-dimensional and temporal autocorrelation being one-directional and one-dimensional. Chapter 4 addresses this space–time data feature by conceptualizing it in terms of spatially structured and spatially unstructured random effects. One theme of that chapter concerns the amount of additional information needed to be able to estimate these two data components with a reasonable degree of accuracy.

Griffith (2008) notes that data forming a short space–time series, one too short to utilize a space–time autoregressive moving average model specification, can be respecified to include a random effects term that contains a spatially structured component in order to account for both serial and spatial autocorrelation. By doing so, space–time heterogeneity can be accounted for in various ways, including specifications involving eigenvector spatial filtering methodology. Chapter 4 contributes to this type of development, recognizing that: (1) components of spatial autocorrelation can vary over time, (2) a spatially unstructured random effects term seeking to account for serial correlation in geographically distributed multiple short time series can better highlight latent spatial autocorrelation, (3) an eigenvector spatial filter can effectively structure a random effects term, and (4) failure to include a spatially structured random effects term can result in biased parameter estimates. Chapter 4 highlights that spatial and temporal autocorrelation interact in a complex way, one capable of generating dramatic over-dispersion from only a fraction of the variation exhibited by a space–time random variable.

10.5 The Spatial Weights Matrix and Eigenvector Spatial Filtering

The spatial weights matrix is a fundamental concept of spatial statistics/econometrics. It characterizes how data values are tied together in overlapping subsets within a given geographic landscape (or a three-dimensional cube for space–time data). It is the source of spatial autocorrelation measurement, establishing which areal unit attribute values are directly spatially correlated. Considerable controversy surrounds these matrices, with debates focusing on, among other topics, sensitivity of outcomes and subjectivity/objectivity and endogeneity/exogeneity of specifications. Chapter 4 addresses a number of features of spatial weights matrices, emphasizing similarity of specifications, and uncovering some guidelines for differentiating between more and less desirable specifications. One of its stated conclusions is that spatial weights matrices merit considerably more study in spatial statistics/econometrics. Another is that eigenvector spatial filters offer the advantage that they set aside considerable noise from a spatial analysis by employing only the data-relevant subset of eigenvectors that can be extracted from a spatial weights matrix.

A spatial weights matrix simultaneously quantifies the relative global strength of a relationship between pairs of locations. Classical statistics assumes that all of these pairwise quantities are zero, whereas spatial statistics/econometrics assumes that at least some are nonnegative. This conceptualization is similar to that for correlated samples in classical statistics and serial correlation in time series. Various spatial weights matrix specifications exist, including: contiguity, continuous distance, nearest neighbor distance, fixed distance band, and spatial interaction. The result is to have a response variable, Y, on both sides of an equation, positing that the y value at a given location is a function of a linear combination of neighboring locations' y values. Initially, spatial weights matrices were binary values based upon contiguity: if two polygons are contiguous, then their spatial weight is 1; otherwise, it is 0. Others include powered inverse distance, negative exponential distance decay, percentage of common boundary, and relative spatial interaction. A common adjustment is to convert matrix elements to a row-standardized form: each row entry is divided by its row sum. This modification recasts the y value at a given location as a function of the arithmetic average of neighboring locations' y values.

An important property of spatial weights matrices, especially considering the controversies surrounding them, is that any reasonable specification (i.e., some mistakes are present) is better than assuming all matrix entries are zero (Griffith and Lagona 1998). Given this caveat, future spatial weights matrix research needs to address three prominent issues: spatial weights matrix comparisons, integrating multiple spatial structures, and space–time specifications (e.g., Chap. 3). Meanwhile, future eigenvector spatial filtering research needs to address how sensitive its results are to spatial weights matrix specifications. One conjecture is that these results are insensitive to the specification because the only difference is that the various sets of eigenvectors render different partitionings of the same mathematical space. A surprising result supporting this conjecture is that the rook and queen definitions of adjacency for a regular square tessellation forming a complete rectangular region yield exactly the same set of eigenvectors (Griffith 2000).

Perhaps changing a specification merely produces either a more or a less parsimonious eigenvector spatial filter.

10.6 Clustering: Spatial Autocorrelation and Location Quotients

Clustering refers to a grouping of items, which can materialize in two conspicuous ways in a geographic landscape: a concentration of some attribute within an individual area unit or a geographic amassing of areal units with similar attribute values. Chapter 6 utilizes the morphism of cluster specifically to analyze location quotients.

Clustering takes on many forms. The location–allocation topic of Chap. 2 alludes to geographic clusters that form in terms of minimizing distance between central facilities and service/demand points. Another is the multivariate technique of cluster analysis (CA). Perhaps no multivariate technique has fascinated spatial analysts more than CA, with its potential for explicitly spatial analysis. An enduring quest among spatial analysts is to define objective methods for comparing areal units to form groups or clusters of these units. CA is something of a reversal of discriminant function analysis (DFA). In DFA, groups are defined a priori by a categorical dependent variable. In contrast, in CA, groups are defined as the outcome of an analysis, with the assignment of observations to clusters based upon one of a number of different algorithms. Consequently, the selected algorithm is of central importance in CA, and the devising of algorithms has occupied a great deal of attention in geography, biology, and other disciplines. A clustering algorithm is some function of the variables with which the clustering is to take place. The decision of which cluster an observation is assigned to is based on a distance measure, in some attribute space, between observations and clusters that already have been formed. At the end of a clustering exercise, the observations within a cluster should be more similar (as measured by the employed clustering algorithm criterion) than observations between clusters. Aside from the selection of a clustering criterion, the key task in CA is deciding upon the number of clusters. To assist in this process, multivariate statistics promotes the use of a visual aid, the dendogram, or tree diagram, coupled with a scree plot.

Chapter 6 applies much of this multivariate conceptualization about clusters to local indices of spatial autocorrelation. Such indices help identify the existence of geographic amassing of areal units with similar attribute values. In doing so supports addressing notions of spatial specialization in a national geographic space, highlighting a very pronounced spatial structure in, for example, Polish employment specialization. Overall, the empirical analysis in Chap. 6 reveals that isomorphisms pertaining to geographic and nongeographic clusters are generic in their aspatial composition and somewhat nonstandard in their spatial composition.

10.7 Spatial Autocorrelation Parameter Estimation for Massively Large Georeferenced Datasets

Spatial autocorrelation analysis started becoming fashionable in the 1970s, with the appearance of Cliff and Ord (1973) and of Paelinck and Klaassen (1979). But in those days datasets often were relatively small, with $n < 100$; for example, Cliff and Ord (1973, pp. 149–152) present power analyses for the Moran Coefficient and the Geary Ratio with $n = 25$, updating this analysis in their Cliff and Ord (1981) book to n between 9 and 81 (pp. 174–178). Continual advancements in computer technology and more widespread georeferenced data collection resulted in n increasing, first to 100s, then to 1000s, and today going beyond 10,000,000s today. Cressie et al. (1996) caution that such statistical techniques devised for data analysis involving n in the 100s or 1000s may well fail when n becomes massive. Levels of significance are a case in point, where virtually every value is statistically significant when $n = 30{,}000{,}000$.

Consider the 7649-by-8581 ($n = 65{,}636{,}069$ pixels; this image is 3.2% larger than the Peruvian image analyzed in Chap. 7) Landsat 7 Enhanced Thematic Mapper Plus (ETM+) complete remotely sensed image of the Florida, USA, Everglades, appearing in Fig. 10.1a, with black denoting the part of the satellite photography that does not pertain to any part of the Earth's surface (and hence can be discarded or ignored). Griffith (2015, p. 2156) describes spatial autoregressive analyses that involve adjusting for regular square tessellations forming such incomplete rectangular regions. The morphism alluded to in Chap. 7 pertains to analytical geometry (i.e., defining and representing areal unit polygons in a computational

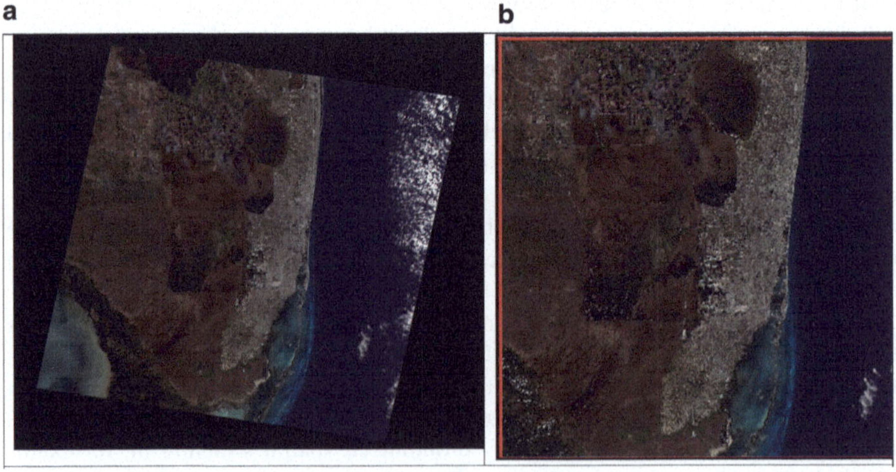

Fig. 10.1 Left (**a**): complete 7649-by-8581 ETM+ remotely sensed image. Right (**b**): 4800-by-5200 central region of the image. These are Landsat 7 Enhanced Thematic Mapper Plus (ETMþ) images of the Florida Everglades for 1 January 2002, downloaded from the webpage http://glovis.usgs.gov/ (last accessed on 21 October 2017); the visualization in (**a**) is a composite image constructed by the author using all spectral bands

way that allows numerical information to be extracted from them) and supports the methodology enabling spatial statistical analysis of these georeferenced data types. For illustrative purposes, consider the central 4800-by-5200 ($n = 24{,}960{,}000$ pixels) complete rectangular region portrayed in Fig. 10.1b.

Two spectral indices were calculated with the B1–B7 spectral band measures from Fig. 10.1b, the normalized difference vegetation index [i.e., NDVI = (B4−B3)/(B4+B3)] and the normalized burn ratio [i.e., NBR = (B4−B7)/(B4+B7)]. Their respective spatial autocorrelation parameter estimates are $\widehat{\rho}_{\text{NDVI}} = 0.958$ and $\widehat{\rho}_{\text{NBR}} = 0.869$, again attesting to maximum likelihood techniques being able to be used with massively large datasets, particularly remotely sensed images. The asymptotic standard error for these estimates is 0.0003; even if they were only 0.001, they would be statistically significant. The respective spatial autocorrelation parameter estimate counterparts for the Peruvian image analyzed in Chap. 7 are about the same for NDVI and substantially greater for NBR (burning actually occurred in Peru; it did not occur in the Everglades).

10.8 Space–Time Data and Semi-saturated Fixed Effects

The topic of Chap. 8 is fixed effects, which are estimated with binary 0-1 (for convenience) variables representing, in geospatial analyses, each location and each point in time as a category in terms of presence (i.e., 1) and absence (i.e., 0). These artificial variates, which create mutually exclusive and collectively exhaustive categorizations, are known as indicator variables in much of the statistics literature and dummy variables in the spatial econometrics literature.[1] In an analysis of variance context, they also are multiplied together in order to evaluate interaction effects (here, between space and time). Including such interaction terms requires more than one attribute value for each space–time combination in order to preserve a sufficient number of degrees of freedom for data analysis purposes. Because a space–time dataset almost never contains more than one measure for each location and point-in-time combination, interaction effects are set aside. This restriction is the source of the chapter's semi-structured fixed effects descriptor.

Fixed effects function as proxies for covariates, hence precluding the inclusion of covariates, and specify that each location and each point in time has a unique arithmetic mean. In doing so, they account for much of the variance in a space–time data series, and they account for heterogeneity in space–time data. They also consume a considerable number of degrees of freedom [i.e., $(n + T)/(nT)$, where n is the number of locations, and T is the number of points in time]. When an indicator variable takes the value of 0, its associated mean disappears from an observation's equational description. Conversely, when an indicator variable takes

[1]They also are known as design variables, Boolean indicators, categorical variables, binary variables, proxies, and qualitative variables.

the value of 1, its associated mean remains in an observation's equational description. Inclusion of these variables allows individual means to be calculated with a single model, pooling all of the data, rather than calculating nT separate means for the individual subgroups of data. The information content of these variables depends on their overall proportion of ones, which is either $1/T$ or $1/n$; these quantities can be viewed as a binomial distribution of T or n Bernoulli trials. The extreme case here of representing all the covariates as indicator variables in a space–time analysis provides a high degree of flexibility in selecting a modeling methodology.

In terms of morphisms, Chap. 8 highlights fixed and random effects in the context of concentration indices, particularly spatial structural indicators.

10.9 Spatial Autocorrelation and Spatial Interaction Gravity Models

Chapters 2–8 analyze Polish employment location quotients. Chapter 9 analyzes the affiliated journey-to-work data, its relevant morphism concerns spatial constraints and geographic clustering of movement in space, and it outlines a relatively new methodology used to account for spatial autocorrelation in gravity model specifications.

Spatial autocorrelation in the geographic distribution of origin and destination attributes and network autocorrelation in the flows between origins and destinations constitute two spatial autocorrelation components contained in spatial interaction data. Chapter 9 highlights that a spatial interaction model specification needs to incorporate both components in order to furnish sound implications about global distance decay. Such models also need to undergo sensitivity analyses in terms of changes in geographic scale and resolution (Griffith and Chun 2015). Chapter 9 pursues this sensitivity challenge by presenting additional infill and increasing domain gravity model estimation results for yet another empirical example.

10.10 Concluding Comments

In summary, corroborating Griffith and Paelinck (2011), this book once again demonstrates that many nonstandard data analytic approaches can populate a spatial analyst's toolbox, with the benefit of improving spatial statistical/econometric analyses. A spatial analyst simply needs to use his/her imagination and the synergisms of multidisciplinary thinking to apply them.

References

Brimberg, J., & Hodgson, J. (2011). Chapter 15: Heuristics for location models. In H. Eiselt & V. Marianov (Eds.), *Foundations of location analysis* (pp. 335–355). Berlin: Springer.

Cliff, A., & Ord, J. (1973). *Spatial autocorrelation*. London: Pion.

Cliff, A., & Ord, J. (1981). *Spatial processes*. London: Pion.

Cressie, N., Olsen, A., & Cook, D. (1996). Massive data sets: Problems and possibilities, with application to environmental monitoring. In The Committee of Applied and Theoretical Statistics (Ed.), *Massive data sets: Proceedings of a workshop* (pp. 115–119). Washington, DC: National Academy Press.

Cressie, N., & Wikle, C. (2011). *Statistics for spatio-temporal data*. New York: Wiley.

Eiselt, H., & Marianov, V. (Eds.). (2011). *Foundations of location analysis*. New York: Springer.

Farahani, R., & Hekmatfar, M. (Eds.). (2009). *Facility location: Concepts, models, algorithms and case studies*. Berlin: Physica-Verlag.

Griffith, D. (2000). Eigenfunction properties and approximations of selected incidence matrices employed in spatial analyses. *Linear Algebra & Its Applications, 321*, 95–112.

Griffith, D. (2008). A comparison of four model specifications for describing small heterogeneous space-time datasets: Sugarcane production in Puerto Rico, 1958/59–1973/74. *Papers in Regional Science, 87*, 341–356.

Griffith, D. (2010a). The Moran Coefficient for non-normal data. *Journal of Statistical Planning and Inference, 140*, 2980–2990.

Griffith, D. (2010b). Modeling spatio-temporal relationships: Retrospect and prospect. *Journal of Geographical Systems, 12*, 111–123.

Griffith, D. (2012). Space, time, and space-time eigenvector filter specifications that account for autocorrelation. *Estadística Española, 54*(177), 7–34.

Griffith, D. (2015). Approximation of Gaussian spatial autoregressive models for massive regular square tessellation data. *International Journal of Geographical Information Science, 29*, 2143–2173.

Griffith, D., & Chun, Y. (2014). An eigenvector spatial filtering contribution to short range regional population forecasting. *Economics and Business Letters, 3*, 208–217.

Griffith, D., & Chun, Y. (2015). Spatial autocorrelation in spatial interactions models: Geographic scale and resolution implications for network resilience and vulnerability. *Networks and Spatial Economics, 15*, 337–365.

Griffith, D., & Paelinck, J. (2007). An equation by any other name is still the same: On spatial statistics and spatial econometrics. *The Annals of Regional Science, 41*, 209–227.

Griffith, D., & Paelinck, J. (2011). *Non-standard spatial statistics and spatial econometrics*. Berlin: Springer.

Griffith, DA., & Lagona, F. (1998). On the quality of likelihood-based estimators in spatial autoregressive models when the data dependence structure is misspecified. *Journal of Statistical Planning and Inference, 69*(1), 153–174

Hansen, P., Mladenović, N., & Taillard, É. (1998). Heuristic solution of the multisource Weber problem as a p-median problem. *Operations Research Letters, 22*(2–3), 55–62.

Kuenne, R., & Soland, R. (1971). Exact and approximate solutions to the multisource Weber problem. *Mathematical Programming, 3*, 193–209.

Paelinck, J., & Klaassen, L. (1979). *Spatial econometrics*. Farnborough: Saxon House.

Part II
Spatial Econometrics

Chapter 11
Introduction to Part II: Spatial Econometrics

This book section summarizes investigations about alternative morphisms (for the algebra of them, see Tsalenko 1990), analyzing how much they can contribute to improving spatial statistic and econometric specifications. Accordingly, a number of remarks are in order.

A first step is again treating isomorphisms, moving from their stricter theoretical definition (i.e., that they are bijective homomorphisms) to a looser definition, to wit specification similarity. Identifying and estimating the **A**-matrix in extended SAR models exemplifies this approach (see Sect. 14.1). A second possible case is that of a semi-isomorphism, in which a weaker aspect of a given specification is isomorphic, although the algebraic expressions might differ, which may be exemplified with spatial discount functions (see Sect. 17.1). The final treatment in this book section is allomorphisms, which are a case of transdisciplinarity, in which specifications from another discipline are considered as possible candidates for spatial econometric analysis. LES functions (see Sect. 16.1), and diversion and curl (see Sect. 14.3), illustrate these cases.

The following discussion considers different applications and then states general conclusions. Implementations of some of the initial in time appear regularly in the presented developments. Over the last years, the authors came across a certain number of possibilities that would allow an enlargement of the scope of spatial statistics and econometrics; these possibilities are specifications from other disciplines or synthetic views of certain spatial econometric tools. This book section presents results of benefitting from these particular morphisms; meanwhile, the authors continue to give attention to other possibilities in the same field. The introduction to each chapter briefly recalls relevant morphism(s).

Permission to reproduce parts of selected results presented earlier elsewhere has been graciously granted by their copyright owners. The authors appreciatively acknowledge these permissions and expresses their sincere thanks for them. One case is that of Sects. 12.3.2, 12.3.3, and 12.6.1, which reproduce system equations and two figures (here Fig. 12.3) from the *Annals of Regional Science* (1999,

D. A. Griffith, J. H. P. Paelinck, *Morphisms for Quantitative Spatial Analysis*,
Advanced Studies in Theoretical and Applied Econometrics 51,
https://doi.org/10.1007/978-3-319-72553-6_11

pp. 573–580); the other concerns Sects. 15.1 and 19.4, for which permission was granted by the editor of *Estadística Española* (Vol. 177/54, pp. 35–52, and Vol. 180/55, pp. 13–32).

The authors also express their thanks to colleagues who helped develop certain parts of this book section about spatial statistics and econometrics; their contributions are duly quoted in the references.

Reference

Tsalenko, M. (1990). Morphism. In M. Hazewinkel (Ed.), *Encyclopedia of mathematics* (Vol. 6). Dordrecht: Springer.

Chapter 12
Tinbergen–Bos Systems: Combining Combinatorial Analysis with Metric Topology

Because theoretical spatial economics form the basis of spatial econometrics, contents of this chapter are largely devoted to that discipline. Tinbergen–Bos Systems (TBS) are a special case of general spatial economic equilibrium modeling; they intend to derive propositions about multi-urban structures generated by the functioning of the spatial economy. Since 1984 ("Tinbergen–Bos Systems Revisited," Kuiper and Paelinck 1984), TBS analysis, which started in the early 1960s, was taken up again by the one author and some of his colleagues, according to a long-term program; results were obtained along the following lines:

- A systematic introduction of input–output relations.
- Empirical checks on the Tinbergen weak and strong hierarchical hypotheses.
- The computation of an upper bound for the number of inadmissible systems.
- Equivalency of transportation cost minimization and profit maximization.
- Introduction of non-discrete metrics, in particular Manhattan and network metrics, on a so-called Manhattan circle.
- The frequency distribution of the network metric over the Manhattan circle.
- The separability of location and allocation aspects in the metricized TBS-version.
- The endogeneity of plant numbers.
- The introduction of prices.

All of these topics are sufficient to qualify as formal extensions; as stated in the title of this chapter; a combination of combinatorial analysis and metric topology is the central morphism, but selected strategic elements of economic theory receive additional, too.

The chapter conclusions give special attention to research directions in spatial econometrics. To keep this chapter self-contained, it presents relevant earlier results, together with citing their sources.

D. A. Griffith, J. H. P. Paelinck, *Morphisms for Quantitative Spatial Analysis*, Advanced Studies in Theoretical and Applied Econometrics 51,
https://doi.org/10.1007/978-3-319-72553-6_12

12.1 Introduction

TBS, developed in the 1960s by their eponymous two scholars (Tinbergen 1961; Bos 1964), were designed to operationalize Lösch's general spatial economic equilibrium model, in particular to derive propositions about the spatial economic landscape in terms of clusters of activities (centers), a specific combination of clusters being called a system. Indeed, if in theory the system of equations characterizing a Lösch equilibrium allows the computing of locations of firms, factory quantities produced, and their unit factory prices, market areas, and boundaries (see Paelinck et al. 1985, Sect. 2.2), no specific landscapes can theoretically be derived. Moreover, the mathematical problem itself in principle is of a nonlinear mixed continuous-discrete nature and consequently rather hard to solve (see Paelinck et al. 1985, for an example). The fundamental idea of TBS is to simplify the Lösch model to a linear one without prices, locations, and market areas, much along the lines of how Leontief operationalized Walras's nonspatial equilibrium.

The original model essentially was based on a discrete (0-1) metric, and though Bos introduced input–output relations, they were not treated in the way the pure final demand model was. In Kuiper and Paelinck (1984), this latter point was taken up again, and Kuiper et al. (1990) introduce a metric on a Manhattan network; in this way, TBS became a special case of a location–allocation problem (Paelinck and Kulkarni 1999).

In what follows, we first expose the main elements of Tinbergen–Bos analysis and its generalization to input–output relations; consideration is given to special center and system structures as well as the problem of inadmissible systems and the nature of the objective function. Next, non-discrete metrics are introduced and the frequency distribution of the implied network metric studied. An important finding here is the separability of location and allocation aspects of TBS. The number of plants in a system is rendered endogenous, and prices are explicitly introduced. The conclusions mention possible spatial econometric work.

12.2 TBS Analysis and First Extensions

As already mentioned in the introduction, a certain number of concepts are central in Tinbergen–Bos theory (see Paelinck et al. 1985, Sect. 2.2). The following is a brief overview of them:

C_1: Every plant producing a given good or service has a standard dimension; so, if total demand for the good or service in question is known, then the total number of plants per activity sector also is known, which short-circuits the principal Löschian problem (i.e., the presence of endogenous binary variables).

C_2: Activities are classified according to the number of plants producing a good or service, with the sector containing the smallest number of plants having the highest rank, numbered 1, and so on.

C_3: Centers are defined as (spatial, local; see Sect. 2.3) clusters of activities, each specific cluster defining a type of center; as hinted at, the precise location provisionally is considered irrelevant, a center being assigned the rank of the highest ranking activity present in it.

C_4: Combinations of different types of centers (subsequently, a given reference space) are called systems.

12.2.1 Input–Output Relations (Kuiper and Paelinck 1984)

Original Tinbergen–Bos analysis was conducted on pure final demand relations (although Bos, as previously mentioned, presented an exercise that included input–output relations). Input–output relations require reasoning in terms of production levels rather than incomes, but the link between the two can be established easily, as the following equation shows for good i:

$$m_i\, y_i = q_i \tag{12.1}$$

where y denotes income, q denotes production, and m denotes the production–income ratio (the inverse of the value added share). If we now consider a problem with two activities and agriculture (the latter being given symbol 0), three types of centers can be constructed: centers with only activities 1, centers with only activities 2, and mixed centers (here given index 3). Figure 12.1 shows how complex the model already is; we subsequently return to this complexity (see Sect. 12.2.2). In Fig. 12.1, the $a_k y_l$ flows represent final deliveries, and the v_2-flows deliver intermediate goods and services between centers (or with agriculture); the connecting lines complete the interpretation.

The solution to the input–output problem rests on a central hypothesis—which subsequently is readdressed (see Sect. 12.2)—that the overall system generated minimizes transportation costs for all of the products transported, these costs being computed with a discrete metric, which boils down to minimizing total export costs at constant unit (per unit product) transportation costs. Via matrix inversion, the equations of the problem are substituted into the objective function, in which appears the central exogenous variable, y^*, total reference area income. Defining

$$\mu_k = n^k / n_k \tag{12.2}$$

where n^k is the number of centers of type k (k = 1, 2; type 3, a mixed center comprising both activities, is treated residually), and n_k is the number of plants in sector k; one obtains the following nonlinear expression for average transportation costs:

$$\frac{t}{y^*} = \frac{A_1\mu_1 + A_2\mu_2 + A_3\mu_1\mu_2 + A_4\mu_1^2\mu_2 + A_5\mu_1\mu_2^2 + A_6\mu_1^2\mu 2_2}{B_1\mu_1 + B_2\mu_2 + B_3\mu_1\mu_2 + B_4\mu_1^2\mu_2 + B_5\mu_1\mu_2^2 + B_6\mu_1^2\mu_2^2} \tag{12.3}$$

where the A_j and B_j are known functions of the parameters of the problem.

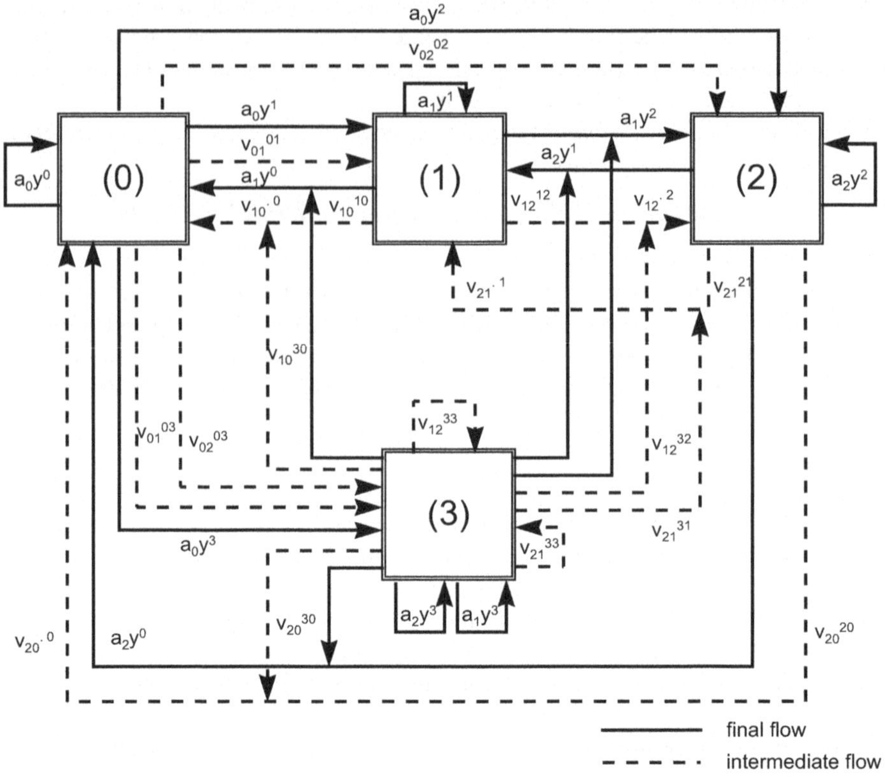

Fig. 12.1 A Tinbergen–Bos system visualized

Various structures of the input–output relations have been investigated, including their absence (which reduces the problem to the original final demand version). In all cases, the solution appears to be the complete type-3 centers. The explanation is not difficult to devise: given the special metric used, transportation costs can be minimized only by setting up mixed centers, which allows saving on interindustry and final deliveries inside urban centers, the sole transportation costs being incurred by movements from and to the agricultural area, but with unit distances. Therefore, an extension to other metrics subsequently is introduced in Sect. 12.3.

12.2.2 Complexity (Paelinck 2000b)

One can reasonably ask the question about how involved the calculations of a TBS become with a growing number of activity sectors. The number of possible systems for I activity sectors, s_I, can be easily computed by combinatorial calculus as

$$s_I = 2^{c_I} - 1 - e_I \tag{12.4}$$

with

$$c_I = 2^I - 1 \tag{12.5}$$

where c_I is the number of possible centers, and e_I denotes excluded or inadmissible systems, the latter being defined as systems not delivering all of the goods and services for which strictly positive demands exist in the reference area. Currently, no analytical expression is known for e_I, but an upper bound can be derived; by constructing three consecutive matrices, a center-system matrix, a center-activity matrix, and a system-activity matrix, one has, in general,

$$e_I < I.s_{I-1} \tag{12.6}$$

and hence the following proposition: e_I/s_I tends toward zero as I increases.

12.2.3 Hierarchy (Paelinck 1995, 1997, Part 1)

Given the potentially large number of possible systems for a given I, Tinbergen proposed some simple criteria that might be verifiable for existing centers and systems.

The first, which can be termed weak Tinbergen hierarchy (WTH), applies to centers, and implies that if a center exhibits the presence of an activity of a certain rank (see the preceding concept C_2), then all activities of lower rank also should be present. The second, which can be termed strong Tinbergen hierarchy (STH), applies to systems, and selects from all possible systems the one that is built up from a complete set of weak hierarchical centers, which obviously number I in types.

A direct proof of the existence of both types of hierarchies requires the use of detailed statistics about the elementary center level (municipalities). By using the following assumptions, and more aggregated material, a statistical approach was possible, the assumptions being

A_1: sector aggregation is over neighboring elementary activities, a not unreasonable assumption in view of homogeneous activity classifications; and,

A_2: center aggregation is over neighboring centers, which again is not unreasonable in terms of common location factors.

Observed statistical tables for this situation should show a quasi-triangular structure, one that could be tested with the help of an appropriate index and an admissible (triangular) frequency distribution for it. In the case of both the Netherlands (1993, 43 sectors and 43 regions) and Japan (1994, 78 sectors and 47 prefectures) the hierarchical hypotheses were not contradicted at confidence levels not exceeding 10–15%. In conclusion, a combined approach by means of location quotients (also see Sects. 1.4, 2.4, 2.5, 3.4, 10.6, and Chap. 6) was advocated; the chapter conclusions return to this point.

12.2.4 *Objective Function*

One can reasonably ask the question about whether systems as such, however defined, tend to minimize global transportation costs, a classical assumption since the advent of early spatial economics. In linear systems, minimization of transportation costs is a necessary condition for profit maximization, which needs to occur independently in each activity sector. Thus, Tinbergen–Bos analysis links to more classical general equilibrium approaches based on the (utility-cum-) profit maximization assumption. This problem is addressed again in Sect. 12.5.

12.3 Metric Extension

As observed at the end of Sect. 12.2.1, discrete metric models tend to generate systems with only complete centers; Sect. 12.2.3 alludes to the statistical possibility of underlying hierarchical structures. The latter fact constitutes a clear invitation to extend Tinbergen–Bos analysis by introducing other metrics, an exercise about which this section reports.

12.3.1 *Manhattan Circles and Distance Frequencies (Kuiper et al. 1990)*

The approach chosen rests on the following three assumptions

A_3: the area to be studied (i.e., the reference geographic landscape) is organized around a rectangular network with unit meshes;

A_4: the area limits are formed by a Manhattan circle with integer radius; and,

A_5: the area is structured by unit Manhattan circles (elementary areas) centered on network nodes, the only candidates for the location of non-agricultural activities; these latter items are uniformly distributed over the elementary areas.

The terminology in assumption A_4 is mathematically incorrect, but has been chosen for the sake of simplicity; to the contrary, that in assumption A_5 is mathematically correct (see the ensuing discussion). Figure 12.2 depicts an example of such a spatial economic landscape with radius $r = 1$; it is a fractal-like picture.

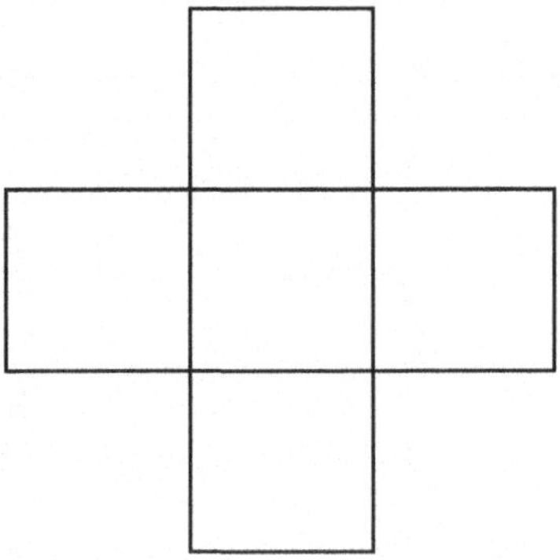

Fig. 12.2 A Manhattan circle with $r = 1$

The number of elementary areas (squares), n_e, as a function of r is given by

$$n_e = 2r\,(r + 1) + 1. \tag{12.7}$$

The square has been chosen as the elementary area with a Manhattan metric ($p = 1$) because for equal surfaces, circles under arbitrary metrics have the smallest average distance. Expression (12.7) has been generalized; it reveals the number of zero distances on a Manhattan circle of radius r, for which we use the symbol $n(r,0)$, and in general $n(r,d)$. The latter expression can be written as

$$n(r,d) = 4\,\left[a_d\, r^2 + b_d\, r + c_d\right] \tag{12.8}$$

the parameters having been computed up to $d = 8$. However, each of these parameters is linked to the previous ones by a recurrence relationship (this property was brought to our attention by J. F. Kaashoek), which means that the complete formulas for any r and d can be derived from the available results.

12.3.2 Equations and Weights

The approach introduced in the preceding subsection can be formalized with the following definitions:

x_{ijk} = the flow of good k between locations i and j;
d_{ik} = the presence or absence of plant k at location i (a binary 0-1 variable);
n_k^* = the (known) number of plants of type k; and,
n^* = the (known) number of possible locations (i.e., $i, j = 1,\ldots,n^*$).

Limiting the present analysis to the case of $k = 1, 2$, the system relations can be written as follows (also see Paelinck and Kulkarni 1999):

$$\sum_i x_{ij1} = 1 \qquad \forall j \tag{12.9}$$

$$\sum_i x_{ij2} = 1 \qquad \forall j \tag{12.10}$$

$$\sum_i x_{ij3} = d_{j1} \qquad \forall j \tag{12.11}$$

$$n_1^* \sum_j x_{ij1} = n^* d_{i1} \qquad \forall j \tag{12.12}$$

$$n_2^* \sum_j x_{ij2} = n^* d_{i2} \qquad \forall i \tag{12.13}$$

$$n_2^* \sum_j x_{ij3} = n_1^* d_{i2} \qquad \forall i \tag{12.14}$$

$$\sum_i d_{i1} = n_1^* \tag{12.15}$$

$$\sum_i d_{i2} = n_2^* \tag{12.16}$$

$$x_{ijk} \geq 0 \qquad \forall i, j, k \tag{12.17}$$

$$d_{IK} \in \mathrm{IN} \qquad \forall i, k \tag{12.18}$$

The composition and location of centers are determined by the minimization of global transportation cost (though this optimization may correspond to profit maximization behavior by individual firms, as stated previously); i.e., of a function

$$f = \mathbf{w}'\mathbf{x} \tag{12.19}$$

where $\mathbf{x}$ is the vector of the x_{ijk}s and $\mathbf{w}$ is a vector depending on the distances between all possible locations i and j (i.e., the d_{ij}s, regardless of the metric used), on unit transportation costs, and on relative quantities shipped (the latter depending on consumption propensities and technical coefficients, as developed subsequently). Budget equilibria imply equality of valued return flows (also addressed in ensuing discussion); transportation costs inside all elementary squares are constant.

Equations (12.9)–(12.11) are demand equations, of products 1 and 2 by agricultural consumers, and of product 2 by consumers working in industry 1; Eqs. (12.12)–(12.14) are the respective supply equations. Equation (12.17) is a non-negativity constraint, and Eq. (12.18) an integral constraint. As such, the problem boils down to a mixed integer-continuous linear program, but as detailed in subsequent discussion, Paelinck and Kulkarni (1999) show it to be separably reducible, for its solution, to a matrix inversion and two sequential continuous linear programs.

Now consider the unit transport cost and the economic coefficients in Eq. (12.19). First, define

a_k = propensity to consume product k;
y^* = total production value (value added in the absence of interindustry relations) of the system (exogenous); and,
$a_k y^*$ = production of sector k.

So, in the absence of interindustry relations, the value transported between sectors k and l equals $a^k a^l y^*$, each firm of sector l demanding $a_k a_l y^* / n_1^*$ from sector k. The total weight for deliveries between k and l becomes $(t_k + t_l) a_k a_l y^*$, so if only two different industrial sectors are considered, as in Sect. 12.2.1, the *relative* weights (excluding distances) relating to the flows x_{ij1}, x_{ij2} and $x_{ij3,}$ the quantities (values) transported between agriculture ($k = 0$) and sector 1, between agriculture and sector 2, and between sectors 1 and 2, become

$$w_1 = (t_0 + t_1) a_0 a_1 n^{*-1} \tag{12.20a}$$

$$w_2 = (t_0 + t_1) a_0 a_2 n^{*-1} \tag{12.20b}$$

$$w_3 = (t_1 + t_2) a_1 a_2 n^{*-1} \tag{12.20c}$$

where the t_k are the unit transportation costs.

In the same vein, Eqs. (12.20a–12.20c) can be generalized to the presence of input–output relations; if a_{kl} is the input coefficient of k in sector l, and m_k is the value added production multiplier of sector k, then for deliveries between sectors k and l, total transportation costs become

$$T_{kl} = [t_k(a_k + a_{kl} m_l) a_l + t_l(a_l + a_{lk} m_k) a_k] y^*. \tag{12.21}$$

Further generalizations, for example, to population interaction and nonhomogeneous linear consumption functions, are possible (see Kuiper et al. 1993, Sects. 12.2.2 and 12.2.3).

12.3.3 Location–Allocation Aspects

Combining the conclusion of Sect. 12.2.2 and the specification of Sect. 12.3.2 indicates that for large I, r, and n_k^*, the solution of the mixed integer-continuous mathematical program implies a heavy computational load, as experience has indeed shown. Applying the substitution of the system equations into the objective function—as was mentioned for the discrete metric case in Sect. 12.2.1; see Eq. (12.9)—results in cross-deliveries—from one elementary area or its node to another—disappearing from the objective function, leaving a simplified system that could be solved by classical linear programming. This particular result was successfully applied by Paelinck and Kulkarni (1999).

To simplify the notation of Eqs. (12.9) through (12.19), consider the following definitions:

$$\mathbf{d} \triangleq [d_1] \quad 5 \times 1 \tag{12.22}$$

$$\mathbf{e} \triangleq [d_{i2}] \quad 5 \times 1 \tag{12.23}$$

$$\mathbf{f} \triangleq [d_{i3}] \quad 5 \times 1 \tag{12.24}$$

$$\mathbf{u} \triangleq [x_{ii1}] \quad 5 \times 1 \tag{12.25}$$

$$\mathbf{v} \triangleq [x_{ii2}] \quad 5 \times 1 \tag{12.26}$$

$$\mathbf{z} \triangleq [x_{ii3}] \quad 5 \times 1 \tag{12.27}$$

Weights can be conformably introduced, the weights for d_{1k}, $k = 1, 2, 3$ being zero because that variable is eliminated. Moreover, for reasons of facility, as $n_1^* = 2$, d_{i1} may be divided into d_{il1}, $l = 1, 2$ (vectors $\mathbf{d_1}$ and $\mathbf{d_2}$), so that that variable becomes binary. The same applies to vector $\mathbf{u}$, with only $\mathbf{u_1}$ appearing in the objective function for obvious reasons, and to vector $\mathbf{z}$. The following function

$$\varphi = \mathbf{a}'\mathbf{d} + \mathbf{b}'\mathbf{e} + \mathbf{c}'\mathbf{f} + \mathbf{g}'\mathbf{u} + \mathbf{h}'\mathbf{v} + \mathbf{k}'\mathbf{z} \tag{12.28}$$

is then to be minimized over the variables (12.22) through (12.27) under the following constraints:

$$\mathbf{i}'\mathbf{d} = 2 \tag{12.29}$$

$$\mathbf{i}'\mathbf{e} = 1 \tag{12.30}$$

$$\mathbf{f} = \mathbf{e} \tag{12.31}$$

$$\mathbf{u_1} = \mathbf{d_1} \tag{12.32a}$$

$$\mathbf{u_2} = \mathbf{d_2} \tag{12.32b}$$

$$\mathbf{v} = \mathbf{e} \tag{12.33}$$
$$\mathbf{f} \geq \mathbf{z}_1 \tag{12.34a}$$
$$\mathbf{f} \geq \mathbf{z}_2 \tag{12.34b}$$
$$\mathbf{z}_1 \geq \mathbf{f} + \mathbf{d}_1 - \mathbf{i} \tag{12.34c}$$
$$\mathbf{z}_2 \geq \mathbf{f} + \mathbf{d}_2 - \mathbf{i} \tag{12.34d}$$

Equations (12.29)–(12.34d) represent a classical case, but a less restricted problem also can be investigated. Equations (12.32a) through (12.34d) rest on the property that if a plant is present in a location, then that location is served first. In fact, given that

$$\mathbf{d}, \mathbf{e}, \mathbf{f}, \mathbf{u}, \mathbf{v}, \mathbf{z} \geq \mathbf{0} \tag{12.35}$$

other constraints are superfluous, due to the constraints (12.29) through (12.34d).

These developments have to be generalized to Manhattan squares of arbitrary radius and to variable numbers of activities and plants.

12.4 The Endogenous Number of Plants with Economies of Scale and Scope

The endogeneity of the number of plants may be derived from economies of scale and scope, as demonstrated in this section.

12.4.1 Economies of Scale

Now production costs are added to transportation costs. Equation (12.18) is combined with the remark about partitioning d_{ik} into d_{ikl}, the latter now being binary (0,1) variables. Here, d_{ik} is divided into d_{iklm}, these variables being defined as unit modules from which plants (index *ikl*) can be built in different sizes, now with $\sum_{ilm} d_{iklm} = n_k^*$, each module having n^*/n_k^* production capacity if product k is delivered to agriculture, and n_{k*}^*/n_k^* otherwise; in this context, Eq. (12.18) represents a technological constraint.

Total production cost for plant l of type k on site i is defined as

$$c_{ikl} = c_k^* d_{ikl} + \sum_m c_{km}^{**} d_{iklm} \tag{12.36}$$

where

$$d_{ikl} = \sum_m d_{iklm} \tag{12.37}$$

and where the c_{km}^{**} (the fixed cost parameters) are at first decreasing and then increasing (scale economies followed by scale diseconomies); summation over m goes from 1 through n_k^*.

An important additional constraint is

$$d_{iklm} \leq d_{ikl(m-1)} \tag{12.38}$$

as otherwise the most favorable fixed cost modules would be selected. Still another remark is that the first term of the right-hand side of Eq. (12.36), summed over all i and l, amounts to $c_k^* n_k^*$, which for spatially invariant unit variable cost c_k^* would be a constant in function (12.19). This result should be compared to what has been said about transportation costs inside elementary squares [see comments after Eq. (12.19)]. A trade-off between economies of scale and transportation costs now is explicitly introduced, which in turn generates concentration or de-concentration of firms of different sizes.

12.4.2 Economies of Scope

Given Eq. (12.36), economies of scope easily can be introduced; neglecting variable costs, and taking two technically compatible products, k and k^*, one can write the joint fixed costs as

$$c_{i(kk*)l} = \beta_{kk*} \left[\alpha_{kk*} \sum_m \gamma_m d_{iklm} + (1 - \alpha_{kk*}) \gamma_m^* d_{ik*lm} \right] \tag{12.39}$$

with $1 < \beta_{kk*} < 2$ being the kk^*-specific economies of scope parameter (costs are not double, but are larger than for a unique product), α_{kk*} the $kk*$-specific weighting parameter, summation not necessarily taking place over the same number of modules, which might be of different sizes or capacities.

12.5 Non-unit Prices

The pricing mechanism is of critical importance and has interaction effects with geographic aspects of a spatial economy.

12.5.1 Price Definition

To begin, an important preliminary question needs to be answered: which pricing is implied as far as the plants and final consumers are concerned, factory or delivered prices?

In what follows, we consider delivered prices. However, already unit factory prices problems arise, despite the fact that all parameters—consumption propensities and input coefficients—are assumed to be fixed (the Leontief hypothesis). As an example, consider the variable costs in Eqs. (12.36) and (12.37), which now are functions of the unit factory prices of all inputs plus what we call the unit transportation charges (i.e., the unit transportation costs multiplied by the distances travelled by the respective products), but with distance being a function of the location of the plants (purveyors and purchasers) concerned. This problem is treated more fully in the next subsections.

12.5.2 Exogenous Prices

The following three cases are considered in this subsection:

- Fixed (Leontief) coefficients
- Variable coefficients
- Economies of scale and scope

12.5.2.1 Fixed Coefficients

If delivered prices are considered, and because coefficients a_k and a_{kl} are expressed in monetary terms, a problem arises from the type of freight pricing (freight rates). Indeed, whether pricing is ad volumen or ad valorem makes a difference. If the pricing is ad valorem, expressions (12.20a–12.20c) and (12.21) can be left as stated; however, if the pricing is ad volumen, then the coefficients should be divided by the delivered price—but the latter, as said in Sect. 12.5.1, is the sum of the factory price of the good delivered (in case of a_k and a_{kl}, the price of product k) plus the unit transportation charge as defined in Sect. 12.5.1. As stated in Sect. 12.5.1, this latter charge is a function of the location of supplier(s) and purchaser(s), which means a function of the solution of the problem and which implies iteratively calculating the coefficients mentioned. Section 12.5.2.3 furnishes an explicit composition of the delivered price, namely Eq. (12.40).

12.5.2.2 Variable Coefficients

This pricing situation is a straightforward case of the so-called Predöhl–Alonso problem (see Paelinck and Nijkamp 1975, pp. 104–124), because the a_k and a_{kl} become functions of the (relative) delivered prices, via the consumers' and producers' demand functions. Thus, corrections and iterations are necessary not only in the case of ad volumen freight pricing but also in that of ad valorem pricing. The only instance where this computation requirement is not the case is that of Cobb–Douglas utility and/or production functions with ad valorem pricing; their parameters are known to be constant in monetary terms (Castermans and Paelinck 1965).

If coefficients are variable, they reflect price elasticities and hence the influence thereof on concentration or diffusion in geographical space (for a formal derivation, see Paelinck 2000a).

12.5.2.3 Economies of Scale and Scope

Not only do the conclusions of Sects. 12.5.2.1 and 12.5.2.2 apply, but now the c_k^*s of Eq. (12.36) become endogenous, because they are derived from input costs. To simplify matters, we *assume* in what follows the absence of primary inputs (also see the remark at the end of Sect. 12.3.2) and the exogeneity of investment costs.

Now parameter c_k^* needs to be transformed into

$$c_{igk}^{\#} = \sum_j \sum_l a_{lk} \left(p_{jl} + t_l \alpha_{jl} \delta_{ji} \right) \tag{12.40}$$

supposing that the a_{lk} are expressed in physical terms, the α_{jl} being the relative delivery (allocation) parameters ($\sum_j \alpha_{jl} = 1$) from different purveyors to a plant indexed by *igk*; subsequently, in Sect. 12.5.3, separate prices p_{jgl} are considered. Again, iteration becomes necessary for the previously mentioned reasons.

12.5.3 Endogenous Prices

This section treats the pricing problem along the same lines as in Sect. 12.5.2.3 (the presence of economies of scale and scope).

Grafting the price fixing mechanism onto Eq. (12.40) together with an optimal mark-up, μ_k, yields

$$\mu_k = -(1 + E_{\mathrm{k}})^{-1}, E_{\mathrm{k}} \leq -1, \tag{12.41}$$

where E_k is product k's own (point) price elasticity, provisionally assumed to be equal in all markets; this quantity should be linked to the demand functions considered in Sect. 12.5.2.2 (for details, see Paelinck 1967, pp. 185–188). Moreover, non-negativity of profits should be assumed (a suggestion we owe to Martin

Beckmann, who proposed this conceptualization as a replacement of the Löschian assumption of zero profits):

$$\pi_{igk} \triangleq \left(p_{igk} - c_{igk}^{\#}\right) d_{igk} \geq 0 \qquad (12.42)$$

where $c_{igk}^{\#}$ is taken from expression (12.40).

A second point, which brings the analysis in line with the Löschian approach, is free entry, which is treated in the manner outlined here (Paelinck et al. 1985, pp. 168–173). Define a new binary (0-1) variable e_{pk}, $p = 1,...,n_k^*$ such that

$$e_{kp} \triangleq d_{i1k} \qquad (12.43)$$

the first module of each plant g, the objective function (12.19) now becoming

$$\phi = \mathbf{i}'\mathbf{e} \qquad (12.44)$$

The total number of plants in all sectors, thus, ensuring maximal entry of plants into the spatial economy considered. The weighting of Eq. (12.44) according to relative entry dynamism in each sector can be considered further. Recall that that number of plants is the main non-convexity in the Löschian general spatial economic equilibrium system of equations (again, see Paelinck et al. 1985, pp. 35–52). Some earlier conditions require being revised; this is in the first occurrence of the case of equations of type (12.3), fixed unit demand now being replaced by fixed unit incomes, for example, rendering

$$\sum_{i,g} x_{ijg1} = \alpha_{i1} \qquad (12.45)$$

where only α_{i1} needs to be explicitly defined as the real (price equals one, e.g.), possibly optimal, consumption share for product 1 in region i.

Accordingly, equations of type (12.19) should be replaced by

$$\sum_{j,g} x_{ijg1} = n*^{-1} \sum_{j} \alpha_{j1} \sum_{g} d_{ig1}. \qquad (12.46)$$

Finally, of note is that absolute prices are generated by fixing incomes (they could be endogenized; see Paelinck et al. 1985, pp. 38–52) and modular fixed costs (they also could be endogenized).

12.6 Conclusions

Some results can be summarized about TBS and the STH. Nevertheless, these two topics offer ample opportunity for considerable future research.

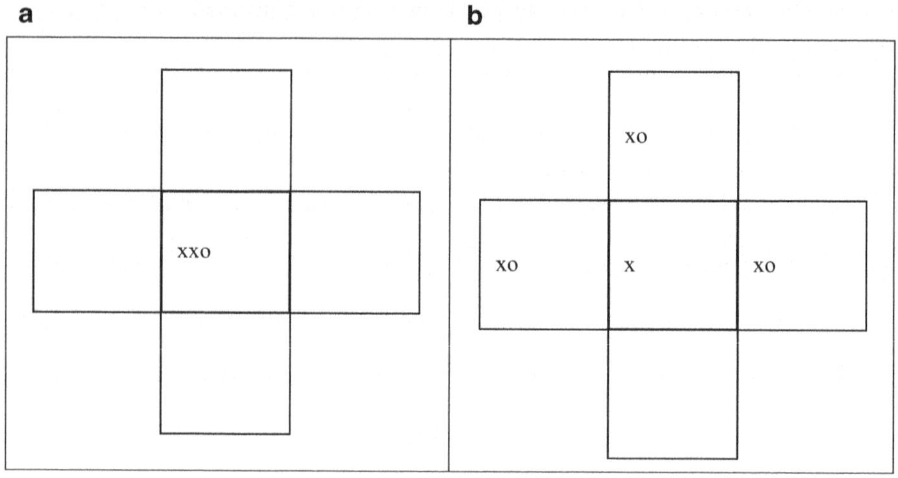

Fig. 12.3 A hypothetical spatial economy landscape. (**a**) Case 1: a possible initial locational pattern. (**b**) Case 2: a static optimum

12.6.1 On Theoretical Spatial Economics

The results obtained by the preceding model refer to static situations that possibly can be compared, in terms of number of plants in the various sectors analyzed or in terms of the effects of varying weights on resulting systems.

In practice, plants, once located, would experience relocation costs, which partly explains their spatial immobility. To take this latter factor into account, and to study the quasi-dynamics of a TBS, past locational situations have to be kept at least during a given time horizon; however, (optimal) flows could be redirected.

Of note is that static and sequential solutions generally diverge. As an example, suppose a starting system like the one portrayed in Fig. 12.3a, the static optimum for a larger number of firms is similar to that portrayed in Fig. 12.3b, where one x-firm and the o-firm of Fig. 12.3a have jumped to an excentric position, whereas in a quasi-dynamic analysis they still would be present in the inner square (equal weights for the two products; figures taken from Paelinck and Kulkarni 1999, with additional examples appearing in Kuiper et al. 1993).

Combining the result obtained by Paelinck and Kulkarni (1999, also see Paelinck 2000a) and the sequential reasoning of this section, systematic histories of spatial quasi-equilibria can be computed now, using information about the evolution of the relevant coefficients constituting the weights in function (12.19).

Geographic outcomes should be compared with observed spatial economic landscapes in order to verify whether or not they could have originated through the processes described by TBS; but this evaluation is a matter for later exercises in spatial econometrics.

Much remains to be investigated in order to better understand the origins and evolutions of existing spatial economic landscapes. Just to mention two additional points: location nodes still are considered as nondimensional points and transportation costs as one-dimensional. Future work should combine potential locations as two-dimensional areas with multi-networks, so as to come nearer to the economic-geographical reality that spatial economists try to underpin with an acceptable explanation.

Finally, the following points need to be addressed in future work:

- Endogenizing freight tariffs (see Lefeber 1958; also see Paelinck and Nijkamp 1975, pp. 426–431)
- Further validation of the Paelinck-Kulkarni (1999) result
- An introduction of primary factors (see Sect. 12.5.2.3)
- Local market behavior (notably the elasticities discussed in Sect. 12.5.3; also see Sarkar et al. 1997)
- Explicit consideration of bilateral flows (a return equilibrium has been assumed throughout the reasoning presented in this chapter)
- Regional policy problems (see Paelinck 2001).

12.6.2 On Spatial Econometrics

Although Sect. 12.2.3 mentions tests of the strong Tinbergen hierarchy (STH) hypothesis that have been conducted, other spatial econometric investigations are in order. To begin, some analytical indicators (Getis and Paelinck 2004) merit being established.

Table 12.1 conforms to what could be called a truncated non-STH case; the integers contained in this table are the inverse rankings of the Centers-Activities column (see Sect. 12.2, point C_2). If the empty cells could be filled, one could compute an entropy measure for the quantities divided by their sum total (here 270); this expression for entropy is given by

$$e = -\sum_i \sum_j x_{ij} \ln \left(x_{ij}\right) \tag{12.47}$$

where i is the activity index, and j is the spatial unit (i.e., center). For the truncated genuine STH table, this value would be 4.2119, whereas for the table as it stands, this quantity is 2.3627 or 0.5610 of the previous value. For different observed tables, this indicator (RE, relative entropy) could be considered to be the endogenous variable one has to explain. Finally, the reason why entropy was selected as an indicator is that linear indicators (e.g., from the numbers appearing in Table 12.1) are more replicable in combinations than are nonlinear functions of them.

But before establishing a non-STH, its reference area should be defined. Little experience is available (in Sect. 12.2.3, two countries were considered as such) to guide doing this task; but some macro-regions (polarized regions; Paelinck 1963) could be preferred over a rather simple choice.

The next problem is that of specifying a suitable model. One possibility is selecting appropriate explanatory variables as regressors in a standard

Table 12.1 A possible 10-activities spatial configuration

	Centers										
Activities	1	2	3	4	5	6	7	8	9	10	# of activities
1	10										1
2		9									1
3			8	8							2
4	7				7	7					3
5		6		6			6				3
6	5		5	5		5		5			5
7	4			4			4		4	4	5
8	3	3			3		3	3		3	6
9	2		2		2	2		2	2	2	7
10	1	1	1		1		1	1	1		7

one-equation model; another specification could be that of an implicit preference model (Hordijk et al. 1976; Ancot et al. 1982), which consists of reversing the usual exercise of optimizing a preference function under constraints, trying to derive certain of its parameters from the observed result of a decision process. Use could be made of certain propositions in location theory (Paelinck et al. 1985, pp. 71–79). The way to spatial econometric system (in the Tinbergen–Bos sense) analysis still needs to be paved; even for the occurrence of STHs (see one in Fig. 12.3b), fully fledged explanations are lacking.

Thinking about this problem could proceed along the following lines. The construction of Table 12.2 assumes that only one activity unit of different types is present in a center, but the givens of a real-world problem are the number of activity units of each type and the number of spatial units. Some activities have a number of units lower than that of the spatial units count, so that they can either join activities of a higher rank or start their own center as the highest ranking activity in it (this is the case of Table 12.1). Other activities exceed in number that of the spatial units; hence, some of them have to join higher ranking activities in larger numbers. Though the process is a dynamic one (Sect. 12.6.1), a static solution could be a starting point for an analysis.

Table 12.2 shows such a result, in which various units of the same activity could be present in each center; this table was constructed using the principle that down and to the left the quantities (units) could not be lower (the solution shown is not unique; one highest rank activity in a center is not a necessary occurrence). The idea is that the concentration of high ranking activities would attract more purveyors (demand effect) or users (supply effect), jointly with externalities of all sorts, and the attraction of lower rank activities also for all sorts of dispersed demands. The result is a rank-size rule (Fig. 12.4), and assuming some proportionality between the number of activity units and the local population, the classical urban rank-size rule would appear.

A first approach to explore assumes that the system in question tends toward maximization of total profits. Accordingly, assume a global profit function $f(\mathbf{e})$,

Table 12.2 A possible truncated STH

Activities	Centers										# of activities
	1	2	3	4	5	6	7	8	9	10	
1	1										1
2	1	1									2
3	2	1	1								4
4	2	2	1	1							6
5	2	2	2	1	1						8
6	3	3	2	2	1	1					12
7	4	3	2	2	2	1	1				15
8	4	4	3	2	2	2	1	1			20
9	5	4	3	3	2	2	2	1	1		23
10	5	4	3	3	2	2	2	1	1	1	25
11	6	4	3	3	3	2	2	2	1	1	27
12	6	5	3	3	3	3	2	2	2	1	30
13	7	6	4	3	3	3	3	2	2	2	35
14	8	7	5	4	3	3	3	3	2	2	40
15	9	8	7	6	5	4	4	3	3	3	52
# of activities	67	54	39	33	27	23	20	15	12	10	300

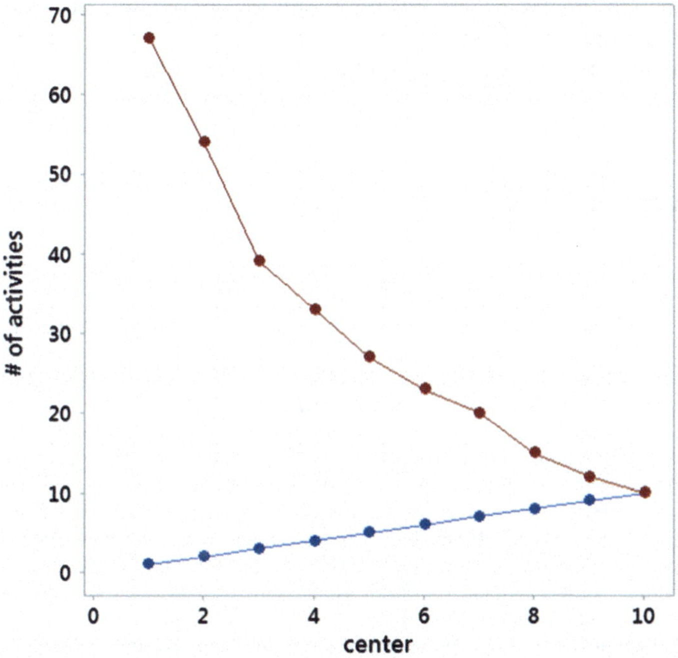

Fig. 12.4 Rank-size graph resulting from a truncated STH. The red dots denote the center totals in Table 12.2. The blue dots denote the centers in Table 12.1 (based on the diagonal entries of this table).

where **e** is the vector of total spatial units entropy (corresponding to the last row of Table 12.2); then assuming that $f(\mathbf{e})$ has the required properties, the following mathematical program specification could be posited:

$$\max_{\mathbf{e}} f(\mathbf{e}) \tag{12.48a}$$

s.t.

$$\mathbf{i}'\mathbf{e} = \mathbf{e}^* \tag{12.48b}$$

where **e*** is the total STH entropy; the result is

$$f'_j(\mathbf{e}) = \lambda, \quad \forall j \tag{12.49}$$

where λ is the Lagrange parameter belonging to Eq. (12.48b); this expression can be rewritten as

$$E_j \propto e_j, \quad \forall j \tag{12.50}$$

where E_j is the profit-entropy elasticity of spatial unit j, which for Table 12.2 leads to the vector [0.8819;0.7580;0.5630;0.4241;0.3674;0.3213;0.2494;0.2033;0.1699]. The accompanying spatial econometric exercise would imply deriving the function $f(\mathbf{e})$ explicitly from a firms' profit maximizing conditions, implying demand functions, production functions, cost functions, and transportation and communication rates, a.o. (see Kau et al. 1986).

Consequently, the last word has not been said about TBS … .

References

Ancot, J.-P., Hughes Hallett, A., & Paelinck, J. H. P. (1982). The determination of implicit preferences. *European Economic Review, 18*(2), 267–289.

Bos, H. (1964). *The spatial dispersion of economic activity*. Rotterdam: Rotterdam University Press.

Castermans, M., & Paelinck, J. (1965). Deux problèmes dans l'analyse d'entrée et de sortie. *Revue belge de Statistique et de recherche opérationnelle, 5*(1), 3–14.

Getis, A., & Paelinck, J. (2004). On analytical descriptions of geographical patterns. *L'Espace Géographique, 1*, 61–68.

Hordijk, L., Mastenbroek, P., & Paelinck, J. (1976). Contributions récentes à l'étude empirique de fonctions de préférence collectives. *Revue d'économie politique, 86*(4), 505–534.

Kau, J., Lee, C., & Sirmans, C. (1986). *Urban econometrics: Model developments and empirical results, Research in urban economics series* (Vol. 6). Greenwich, CT: JAI Press.

Kuiper, F., Kuiper, J., & Paelinck, J. (1993). Tinbergen-Bos metricised systems: Some further results. *Urban Studies, 30*(10), 1745–1761.

Kuiper, J., & Paelinck, J. (1984). Tinbergen-Bos systems revisited. In J. Pillu & R. Guesnerie (Eds.), *Modèles Economiques de la Localisation et des Transports* (pp. 117–140). Paris: E.N.C. P..

Kuiper, J., Paelinck, J., & Rosing, K. (1990). Transport flows in Tinbergen-Bos systems. In K. Peschel (Ed.), *Infrastructure and the space-economy* (pp. 29–52). Hiedelberg: Springer.

Lefeber, L. (1958). *Allocation in space*. Amsterdam: North Holland.
Paelinck, J. (1963). La teoría del desarrollo regional polarizado, Revista Latino Americana de Economía (Caracas), no. 9: 175 229 [reproduced in Cahiers de l'ISEA, Série L (Economies Régionales), mars 1965, pp. 5 47, a: "La théorie du développement régional polarisé", and in Italiea as: "La teoria dello sviluppo regionale polarizzato", in A. Testi (a cura di), Sviluppo e pianificazione regionale, Giulio Einaudi, Torino, 1977].
Paelinck, J. (1967). *Analyse micro-economique*. Unpublished manuscript (mimeographed). Faculté des Sciences Economiques, Centre de Recherches, Namur University, Namur, Belgium.
Paelinck, J. (1995). *Four studies in theoretical spatial economics: Part 1, Empirical evidence on Tinbergen-Bos systems*. Working Paper Series No. 100. Munich: University of Munich, Centre for Economic Studies.
Paelinck, J. (1997). Two studies on Tinbergen-Bos systems: Part 1, An application of Tinbergen-Bos analysis to the case of the Japanese prefectures; Part 2, Frequency distribution of distances on a Manhattan circle. *Oikonomika, 34*(1), 1–22.
Paelinck, J. (2000a). On aggregation in spatial econometric modelling. *Journal of Geographical Systems, 2*(2), 157–165.
Paelinck, J. (2000b). *Controlling complexity in spatial modelling*. Working Paper No. 362. Munich: CESifo Group.
Paelinck, J. (2001). Regional competition in the framework of quasi-dynamic Tinbergen-Bos systems. In J. Roy & W. Schulz (Eds.), *Theory of regional competition* (pp. 80–87). Baden-Baden: Nomos-Verlag.
Paelinck, J., Ancot, J.-P., Gravesteijn, H., Kuiper, J., & ten Raa, T. (1985). *Eléments d'Analyse Economique Spatiale*. Paris: Anthropos.
Paelinck, J., & Kulkarni, R. (1999). Location-allocation aspects of Tinbergen-Bos systems. *The Annals of Regional Science, 33*(4), 573–580.
Paelinck, J., & Nijkamp, P. (1975). *Operational theory and method in regional economics*. Farnborough: Saxon House.
Sarkar, J., Gupta, B., & Pal, D. (1997). Location equilibrium for Cournot oligopoly in spatially separated markets. *Journal of Regional Science, 37*(2), 195–212.
Tinbergen, J. (1961). The spatial dispersion of production: A hypothesis. *Zweizerische Zeitschrift für Volkswirtschaft und Statistik, 97*(4), 412–419.

Chapter 13
Time, Space, or Econotimespace?

Time has been analyzed over and over again by a wide range of scholars, much more than space has, which came quite later to the fore. In this chapter, we underscore a certain number of joint aspects of these two dimensions, aspects that are relevant to and of interest for the coming chapters. Contextualizing references—not quoted in the study hereafter, but consulted—are Asada and Ishikawa (2007), Barker (1996), Corpataux and Crevoisier (2007), Lesse (1983), Maddison (1995), and Melvin (1990). As an aside, the approach presented here is specific to the present field of research, as a reader may well notice.

The formal contributions made by the contents of this chapter is the triad mentioned in the title; some econotimespace models are developed, showing, as was done in the previous chapter, that a close integration of space–time and economic theory could lead to new insights in economic decision making.

13.1 A Conceptual Analysis

First a number of fundamental issues require consideration.

13.1.1 Time

Time has been considered since times immemorial (e.g., see Pucelle 1955), and many dimensions of it have been analyzed. This is not the narrative to go into these details nor to dwell on sayings like "if I think about time I know exactly what it is, but if you ask me to define it I am unable to do so" (Augustinus is said to have declared this). Reflections on physical time (Barreau 2005; Brans et al. 1988; Couderc 1954; Greene 2004; Hawking 1989; Prigogine and Stengers 1988; Rovelli

D. A. Griffith, J. H. P. Paelinck, *Morphisms for Quantitative Spatial Analysis*,
Advanced Studies in Theoretical and Applied Econometrics 51,
https://doi.org/10.1007/978-3-319-72553-6_13

2008; Varii Auctores 2000) are interesting by themselves, but economics is about behavior of human beings, not about elementary particles.

There are a certain number of time aspects that should be considered here from the economic point of view. The first, and probably the most important one, is that time is an economic good or service, has value by itself, and can be traded at a certain price ("time is money," indeed . . .). Examples are numerous (e.g., working hours, money lent over a certain period), but an additional aspect to it exists: in all cases, time has a certain quality (one hour's time of an unskilled worker has a different price than the same hour devoted to a scientific lecture, not to mention a manager's prime time; time is valued differently according to the solvency adjudged to a borrower; subprime loans are here to testify to this . . .). This breadth makes for a fantastically varied market for all sorts of times, all of them being marketable; time *is* an endogenous variable!

A second aspect of time is that we can picture the future, which we also can plan in our strategies, strategies based or not on rational behavior (Sheffrin 1996). This feature implies that our economic decisions—and the resulting variables—are a function not only of (present and) past values but also of future ones. Physical systems do not look into their future, whereas economic ones do, and if physical models can lead to chaotic outcomes, they are only the result of the way their variables interact according to time's arrow, pointing from the past to the future (Gould 1987). In mathematical terms, the transition matrix of those systems is lower (block) triangular, the upper (block) triangular part missing. In ideal economic models, where prospective values of some variables play—or should play—a crucial role, this is *not* the case.

Reflecting on the contents and messages of this overview, the "end of time" (Barbour 2000) is far from coming up soon!

13.1.2 Space

As already stated, space attracted attention much later in economic analysis than did time, but it enjoyed the advantage of having a large amount of literature to rest upon, to mention only geometry and, later, topology (an interesting reference is Gray 1979).

Before studying its place in economic analysis, a most important question should be posed: what is the space we are talking about? Space is a very precise mathematical concept (Paelinck et al. 1983, pp. 37–52) whose meaning should be recalled explicitly here.

The mathematical definition of time rests on that of a *structure*, a set consisting of two subsets, one of elements and the other of relations between these elements, such that

$$S = \{O, R\}, \tag{13.1}$$

where S stands for structure, O for the set of elements, and R for that of relations. In spatial economic analysis, O could be a set of relevant geographical sites, and R a one-element set, the distance relation. This conceptualization turns the set O into a *space*—a metric space, in this case—of which the elements are now termed *points*. Adding some characteristics to each of the sites yields a pre-geographical space (Beguin and Thisse 1979). Another structure is that of measure spaces (Fadden 1977).

No economic analysis can be performed at this point because economic relations still are absent, relations that could set in motion economic activity. Such relations involve preference and productivity, generating choice and production spaces, leading to the possibility of defining and computing *spatial economic allocations* (SEAs), the assignment of goods and services to agents, and the three to sites.

This backdrop raises a first problem: how should the set of sites be defined? Quite a few analyses use a one- or bi-dimensional continuous set of sites, but, in reality, relevant sites form a discontinuous set, introducing a crucial element into spatial economic analysis, to wit non-convexities, probably the worst worry when trying to solve for SEAs. A classical example of this is Lösch's general spatial equilibrium model, in which a—provisionally unknown—*number* of production plants have to be assigned to their optimal locations on a continuous set of possible sites (for details, see Paelinck et al. 1983, pp. 127–142). Interestingly, this number problem appears in another earlier context (von Franz 1978); it characterizes time, too, in which discontinuous sets of periods may be considered. But this is not the end of our spatial worries.

Undeniably, postulating a metric space is of the essence in spatial economics; but then doing so raises the question of the choice of an appropriate metric: should it be a Euclidean one, a Manhattan one, or some Minkowski p-metric one? And there are thousands of others; one would be glad to have such a simple choice, were it not that results (propositions) in spatial economics are not indifferent to the metric chosen. A famous example is that the optimal location in a Weberian problem depends on the metric chosen: it is unique with a Euclidean one, or it can be a one-dimensional infinity with a Manhattan one (Huriot and Perreur 1990). But the use of any metric should be justified for the problem at hand (it would be, for sure, for a Manhattan metric); metrics matter for the solution of spatial problems. As can be seen i.a. in Chap. 16, the use of a metric often simplifies a problem, because it sort of reduces the relevant universe to a one-dimensional one.

A further remark is in order here: a metric is a nondimensional entity, in the sense of *dimensional analysis* (de Jong 1967). Thus, distances cannot be bridged, but lengths can be travelled on dimensional roads. Homogeneity of equations is an important formal requirement and should be carefully checked, as is done in Sect. 13.1.3 hereafter.

One last remark: a metric is only one element of the structure of spatial economic problems. As mentioned previously, next to metric spaces are algebraic spaces, and hence the use of adequate algebra also needs to be checked. Just starting with associativity and distributivity can lead to inappropriate models; other algebraic spaces (e.g., a min-algebra; see Paelinck 2000) could be more advisable.

Nevertheless, next, we address our fundamental problem.

13.1.3 Space–Time, Rather Than Just Space or Time?

In restricted relativistic physics, a central equation is that of the invariant interval

$$s_{10}^2 = c^2(t_1 - t_0)^2 - l^2\left[(x_1 - x_0)^2, (y_1 - y_0)^2, (z_1 - z_0)^2\right], \tag{13.2}$$

in three dimensions, with c denoting the speed of light, t time, and x, y and z shifting time and space coordinates. Does an analogous space–time relation exist in spatial economics?

Yes, it does; however, the argument is very different from the physics' one (which hints that Clifford algebra is probably not immediately applicable to spatial economic problems; see Hestenes 1966). Suppose a length l^* has to be travelled; this only can be done at a cost. This cost comprises the following two components:

- The cost of time lost during travel; assume this to be a convex function of time itself, $c_t(t)$.
- The cost of travelling a given distance at a certain speed, this unit speed cost increasing with velocity (remember the Concord experience!); if this function is $c_v(l^*/t)$ with $c_v' > 0$, then it is convex in time.

Therefore, total travel cost is given by

$$c\sim = l^* c_v(l^*/t) + c_t(t), \tag{13.3}$$

with $c\sim$ again being convex, its minimum with respect to time (because minimizing is what a rational economic agent will do!) being given by the solution to

$$c\sim' = -c_v' l*^2 t^{-2} + c_t' = 0. \tag{13.4}$$

Hence,

$$\overset{\circ}{t} = l^* \sqrt{c_v'}/c_t', \tag{13.5}$$

an elegant formula that brings together an element of topological space, l^*, and time; the Pareto-circlet ^{o} stands for the optimal value. Therefore, the higher the marginal speed cost, the more time one invests in a trip, and vice versa for the marginal cost of time, a logical outcome. Expression (13.5) is dimensionally correct, in the sense of dimensional analysis (again, see de Jong 1967).

As an example, consider $c_v = c_{vo}(l^*/t)$ and $c_t = c_{to}t$; then the result is $t^o = l^* \sqrt{c_{vo}}/c_{to}$, so that time is proportional to length, but only in this case, not in the general case of Eq. (13.5), because the functions under the square root sign still are functions of time. This finding invites a closer study of the functions c_v and c_t.

But not only the producer connects space and time, so also does the consumer; this latter connections is exemplified next.

Let $u = u[a(t, l), b]$ be a consumer's utility function; a denotes some activity, of which the utility is increased by duration of it and the distance traveled (e.g., holidays), and b are all other activities; constrains are on income, r^*, and available time, t^*, such that

$$p_a(l)t + p_b b \leq r^*, \tag{13.6a}$$
$$t \leq t^*. \tag{13.6b}$$

Assuming u to be concave in its arguments, maximizing u leads to

$$u'_a v'_t - \mu - \lambda p_a(l) = 0, \tag{13.7a}$$
$$u'_a v'_l - \lambda p_a(l)t = 0 \tag{13.7b}$$

and assuming $\mu = 0$,

$$l = \left(v'_t/v'_l\right)E(p_a, l)t, \tag{13.8}$$

where E is the price-distance (length) elasticity, Eq. (13.8) again being dimensionally homogeneous.

Therefore, in spatial economics, time and distance—read: length—are interconnected, but via typical economic behavior. In other words, time creeps up systematically, but in different forms of relations, in spatial economic analysis, as well as, later on, in spatial econometrics. Although the formula looks continuous, in practice, economic agents think in terms of slices of geographical space and operating time and hence non-convexities (also see Paelinck and Mur 2013).

13.1.4 Toward Spatial Econometrics

In what follows, some econometric implications will be considered; to avoid confusion, we will speak about space as econo-space, meaning that all the relevant spaces, as discussed earlier, will be present in the analysis. The most important of them have been mentioned, topological and economic space (production and consumption), and also the importance of an appropriate algebraic space; for other, mostly implicitly admitted, spaces (measure space, probability space ,. . .), one is again referred to Paelinck et al. (1983, pp. 37–52).

13.2 Space–Time Spatial Econometrics

This section addresses the following two problems in spatial econometrics: the link between space and time as considered in Sect. 13.1.3 and the possible resulting misspecifications of the equations of a spatial econometric model.

13.2.1 Space–Time Relations

Most (spatial) econometric applications refer to aggregated quantities resulting from a large number of decisions, all of them impregnated by an economic unit's underlying idiosyncrasies (see the parameters presented in Sect. 13.1.2). Accordingly, a way of capturing space–time relations in spatial econometrics needs to be devised.

A possible solution route is to exploit the (marginal) rate of substitution, encountered in demand and supply analysis. Consider a function *f(z)*; the following expression:

$$df = f'_x dx + f_t dt = 0, \tag{13.9}$$

pictures a change of variables x and t that leaves the value of f unchanged. The following expression:

$$r \triangleq |f_x / f_t| \tag{13.10}$$

is called the marginal rate of substitution; absolute values have been considered because, as in spatial econometrics, negative partials can appear. If the function f is linear, Eq. (13.10) reduces to $r = a/b$, these parameters corresponding to the partials in Eq. (13.10).

A first exercise has been conducted with the quantities appearing in Tables 14.9 and 14.10 from Griffith and Paelinck 2011, Chap. 14. It is a special case, to wit one of composite parameters; i.e., a parameter is composed of a generic part, a region-specific one and a time-specific one, in this case a parameter $c = c^* + a_r + b_t$, and to be a proxy to r, the ratio a_r/b_t has been chosen. Because the quantities appearing in these tables are disaggregated, they were aggregated for this exercise; moreover, because b_t was disaggregated on a yearly basis, its average value over the period was taken. Table 13.1 shows the values of the pseudo-rates of substitution; the order of the regions is detailed in Table 15.1. The values appearing in this table are highly

Table 13.1 Reaggregated data for 11 Belgian regions

Region	Pseudo-rate of substitution
1	834.0974
2	694.8073
3	166.1805
4	13114.44
5	759.8413
6	800.3848
7	31.21704
8	982.5475
9	858.4868
10	529.1256
11	1045.741

Table 13.2 Pseudo-rates of substitution for the four Dutch macro-regions

Macro-region	North	East	South	West
North	1	0.269864	0.018806	0.566526
East	0.131842	1	0.330194	0.011536
South	0.941519	0.056714	1	0.551767
West	4.365841	3.990399	1.480546	1
North	1	0.002014	0.099547	0.291467
East	726.6875	1	26.5	148.125
South	1.214623	2.191038	1	2.273585
West	3.452326	1.324419	0.275581	1

variable and rather large, the latter feature being due to the weakness of the b_t coefficients.

The next exercise is a real rate of substitution one, the quantities being taken from Table 14.3 of subsequent Chap. 14, this chapter giving information about the origin of two tables. Table 13.2 gives the results for the four Dutch macro-regions (see also Fig. 14.1).

Once more, the sizeable variability of the parameters is apparent; only one has an exceedingly large value, for the reason previously mentioned.

13.2.2 Space and Time Misspecification in Spatial Econometrics

This subsection addresses the problem of selecting a dynamic process underlying a (multiple) time series. The use of a flexible discount function reveals itself to be a precious aid in selecting an appropriate specification.

13.2.2.1 The Problem

Because spatial econometrics is about economic behavior, special attention has to be given to the *specification* of an underlying spatial econometric model or the first moments of some distributions, so to speak. This important point is introduced next, so as to set the stage for content presented in this subsection.

The standard model in spatial econometrics can be said to be the spatial lag model, which may be written as

$$\mathbf{y} = \rho \mathbf{W} \mathbf{y} + \boldsymbol{\varepsilon}, \tag{13.11}$$

where **W** denotes a spatial weights matrix and ρ a spatial lag parameter. Equation (13.11) is a (time) static model. This feature raises the question asking how a static model is generated from inherently dynamic (causal; on their links, see Heyde 1957 and Casti 1997, pp. 189 a.f.) behavior. As an example to illustrate this point, assume the following simple one-equation one-lag model:

$$y_t = ax_{t-1} + \varepsilon_t, \tag{13.12}$$

and further assume that observations are only available for a double period, so that

$$y_t^* = y_{t+1} + y_t, \tag{13.13a}$$
$$= ax_t^* + a(x_{t-1} - x_{t+1}) + (\varepsilon_t + \varepsilon_{t+1}), \tag{13.13b}$$

where $x_t^* = x_{t+1} + x_t$; this model is supposed to be fitted as a static one. If the error term has the usual properties, its sum would have zero expectation and double variance (discussion about the influence of spatial aggregation on the error terms, see Paelinck 2000; and about the influence of dynamic misspecification, see Balestra 1982). But a residual term exists in the x-variable, and assumptions should be made about that two-period difference. This difference could be a constant or a constant plus a stochastic term; either a constant rate of growth or a random walk could exist. Nevertheless, dynamics cannot be ignored and have to be specified. Maybe the problem could be neglected if the aggregation process runs like

$$y_t^* = \lambda y_{t+1} = (1 - \lambda)y_t, \tag{13.14}$$

with λ close to 1, in which case the second term of the right-hand side of Eq. (13.13b) becomes $a(1 - \lambda)(x_{t-1} - x_{t+1})$, which might produce only a negligible bias in the estimation of parameter a. Anyhow, caution is to be advocated, and this is the reason why the next section deals with the topic of space–time specifications in more detail.

As for aggregation over space, it has been examined extensively in Paelinck (2000), referring to a problem studied in particular by spatial statisticians as the Modifiable Areal Unit Problem (MAUP), the possible use of territorial units of different sizes and zonations. In a genuine econometric spirit, this surface partitioning exercise can be treated as a spatial aggregation problem, producing some disturbing consequences for a spatial econometrician. One of these consequences may be summarized as follows (from the previously cited article): "The important result is that in general econometric aggregation, if only one macro-aggregate is considered, *just one* parameter bias is present in the macro-model; in meso-aggregation, as it took place here, *every meso-area* has its own specific aggregation bias, which leads to parameter variability between meso-areas, and this might result, in econometric estimation, in some sort of (biased) average value, depending on the characteristics of the sample being investigated and the particular spatial aggregation specification."

For larger models, the implicit bias is even more complex; moreover, the stochastic terms of a model reveal heteroscedasticity and spatial autocorrelation under very general conditions. Of note here is that resulting conclusions impose the use of appropriate specifications adapted to each problem at hand; a possible technique for achieving this end is that of composite parameters—at least when the number of degrees of freedom is adequate—in order to take account of the *specific* bias inherent in each meso-economic spatial unit included in a cross-sectional analysis (Ancot et al. 1978). But then, how should one sort out spatial heterogeneity and spatial bias? Recently, filtering data for observational errors, and then for spatial aggregation bias, was proposed in Griffith and Paelinck (2011, Chap. 18); their method was applied to a series with maximal spatial complexity, after which complexity was reduced by two-thirds, allowing a simple linear model to be fitted to the filtered data.

13.2.2.2 Specifications

Several model specifications will be analyzed hereafter from the aggregation point-of-view.

Only One Exogenous Variables

Starting with a single exogenous variable model such as Eq. (13.12), assume that the true specification is

$$y_t = \lambda a x_{t-1} + (1 - \lambda) a x_{t-2}, \tag{13.15}$$

which means a shared overlap of the impulse over two partial time periods. Assuming $x_t = \alpha x_{t-1}$ yields

$$y_t = a\left[\lambda + \alpha^{-1}(1 - \lambda)\right] x_{t-1}, \tag{13.16}$$

over- or under-estimating a depending on the sign and magnitude of α if model (13.16) is substituted for model (13.15). If a static model (x_t denotes the exogenous variable) is used, the result would be

$$y_t = \alpha^{-1} a\left[\lambda + \alpha^{-1}(1 - \lambda)\right] x_t, \tag{13.17}$$

again with a constant ratio α between successive values of the exogenous variable, leading to the same conclusions as before. If α is known, then a can be identified only if λ also is known, which generally is not the case.

Dynamic Processes

Returning to the problem raised about Eq. (13.11), assume the true model to be

$$y_t = ay_{t-1} + b \tag{13.18}$$

and, again, two subsequent time periods to be aggregated; therefore,

$$y_{t+1} + y_t = a(y_{t-1} + y_{t-2}) + a(y_t + y_{t-2}) + 2b, \tag{13.19a}$$
$$= a^2(y_{t-1} + y_{t-2}) + 2(1 + a)b. \tag{13.19b}$$

In principle, parameters a and b are identifiable if knowledge is available about process (13.18) and the aggregation procedure. But because knowledge about Eq. (13.18) is rather dubious, only result (13.19b) can be obtained.

Now if the underlying process is

$$y_t = ay_{t-1} + a^* y_{t-2} + b \tag{13.20}$$

and one estimates a one-lag model, y_{t-2} can be substituted for by the instant ratio $r_v = y_{t-1}/y_{t-2}$, rendering

$$y_t = \left(a + r_v^{-1} a^*\right) y_{t-1} + b, \tag{13.21}$$

However, the coefficient obtained for the lagged variable [e.g., y_{t-1} in model (13.21)] is variable over the estimation period as a result of varying r_v, and consequently both coefficients a and a^* cannot be identified.

The Multiple Variable Case

Here, space and the spatial weights matrix **W** can be reintroduced, generalizing equation (13.20) to

$$y_t = \rho W y_{t-1} + \rho^* W y_{t-2} + b. \tag{13.22}$$

Estimating only a one-lag model, one obtains

$$y_t = (\rho W + \rho^* W r^{-1}) y_{t-1} + b, \tag{13.23}$$

with $\mathbf{r}^{-1}$ denoting the diagonal matrix generalizing r_v of Eq. (13.21). Here, the static case can be envisaged, yielding

$$y_t = r{*}^{-1}\left(\rho W + \rho^* W r^{-1}\right) y_t + b, \tag{13.24}$$

with an obvious interpretation of $\mathbf{r}^{*-1}$, but in the resulting condensed specification

$$\mathbf{y_t} = \rho^{**} \mathbf{W} \mathbf{y_t} + \mathbf{b}, \tag{13.25}$$

ρ^{**} can be shown to be non-identifiable for row-normalized spatial weights matrix **W**.

13.2.2.3 Simulations

In order to obtain some insight into the order of magnitude of empirical biases, the coefficients of models (13.20) and (13.22) have been estimated, starting from simulated underlying models (13.18) and (13.22).

Models (13.20) and (13.21).
The true model (13.18) has been specified numerically as

$$y_t = 1.1y_{t-1} - 0.15y_{t-2} + 1, \tag{13.26}$$

and simulated with starting points (10, 11). Table 13.3 reports the next ten quantities. Of note is that a negative coefficient has been chosen for the second-order lag. Process (13.26) can be shown to converge to the value 20.

Equation (13.15) was reduced to just a first-order lag process and its parameters estimated from the data appearing in Table 13.1 by the SDLS technique (Griffith and Paelinck, Chap. 12). As progressively less data for the past are available, reduced potentials were computed, but with (equally spaced) linearly decreasing weights, down to 0.1 (sometimes full past data are available; see Agarwal 2009). This procedure produced a lag parameter estimate of 0.9396, and a constant estimate of 1.2022, which corresponds to the result of Eq. (13.20) for $a^* < 0$ and $r_v > 0$.

This analysis continues in Sect. 13.2.2.5.

Table 13.3 Data for simulating equation (13.26)

y_t	Values
1	11.60
2	12.11
3	12.58
4	13.02
5	13.44
6	13.83
7	14.20
8	14.54
9	14.87
10	15.17

Models (13.22) and (13.23)

To implement model (13.22), the spatial weights matrix **W** for two regions has been constructed with the row vectors [0.5, 0.5; 0.1, 0.9]. These numbers resulted in the following equations, with $\rho = 0.8$ and ρ*= 0.25 and constants 0.3 and 0.5:

$$y_{1t} = 0.40y_{1,t-1} + 0.40y_{2,t-1} + 0.125y_{1,t-2} + 0.125y_{2,t-2} + 0.3, \quad (13.27a)$$
$$y_{2t} = 0.08y_{1,t-1} + 0.72y_{2,t-1} + 0.025y_{1,t-2} + 0.225y_{2,t-2} + 0.5. \quad (13.27b)$$

This system of equations has two real negative roots and two complex conjugate roots outside the unit circle; thus, it is not convergent.

Simulations with stating point (10, 11; 5, 5.2) produced in Table 13.4.

Estimating the parameters of specification (13.22)—again with the weighting mentioned in section "Models (13.20) and (13.21)"—resulted in the quantities reported in Table 13.5.

Remarkably, the condensed ρ is 25% larger than the original first-order one, which is in line with the results shown by Eq. (13.22); again, Sect. 13.2.2.5 presents a continuation of this analysis.

13.2.2.4 Solutions

Various suggestions could be made that neutralize time and space biases present in spatial econometric work. One possibility is to test, if possible given the data, several underlying specifications against each other, but the alternatives are rather numerous. Another way of attacking this problem is to introduce time and space

Table 13.4 Results of simulating equations (13.5)

y_{1t}, y_{2t}	Values of y_{1t}	Values of y_{2t}
1	8.66	6.50
2	839	7.32
3	8.48	8.12
4	8.90	8.88
5	9.49	9.64
6	10.17	10.42
7	10.93	11.22
8	11.74	12.06
9	12.59	12.92
10	13.48	13.81

Table 13.5 Parameter results for Eq. (13.22)

Parameters	Variable y_1	Variable y_2
Condensed ρ	1.0414	1.0414
Constant	−0.0153	0.1815

flexible discount functions; this section summarizes some ideas about this possibility.

The shape of a spatial or temporal lag function can be quite sophisticated; the spatial weights matrix **W** of Eq. (13.21), normalized or not, often can be considered an oversimplification of spatial interaction effects, and the same applies to time interaction. The reason is that distances, which are often implied, do not necessarily mean physical ones, but mostly refer to functional distances [e.g., economic structures (Kocornik-Mina 2007)]. Flexibility in this case can be introduced using some appropriate functional form, such as the well-known one-parameter Poisson (distribution) function $f(n) = e^{-\mu}\mu^n/n!$, where n is the time-lag degree. Use of this latter function in time series has shown that a first-order lag is not necessarily dominant (Agarwal 2009). The same applies to spatial lags; different possibilities exist for introducing flexibility, such as the use of a so-called Tanner function, which may be specified as follows:

$$f(d,\gamma) = \gamma^{*} d e^{-\gamma d}, \tag{13.28}$$

with d an appropriate distance function, $\gamma \geq 0$, and γ^* the usual normalizing constant (γ^{-2}). Another possibility is the use of the following function:

$$f(d,\gamma) = e^{(1-\gamma^{**})}[\ln(1+\gamma) + \gamma^{**}](1+\gamma d)^{-1}, \tag{13.29}$$

again with $\gamma \geq 0$. Both functions allow for a maximum effect, Eq. (13.29) showing it at a distance $d^* = (e^{(1-\gamma^{**})} - 1)/\gamma$, such that for $\gamma = 0$, the maximum effect operates at $d = \infty$, and for $\gamma = \infty$ at $d = 0$ (further details are available in Ancot and Paelinck 1983).

Data sometimes allow the computing of intermingled space and time reactions (e.g., panel data); how space and time have been combined is illustrated here by way of a bivariate statistical approach. The specification problem can be approached as outlined here (Griffith and Paelinck 2009). The degree of contiguity was selected to be a distance function and years to be time units. This implies that a discrete bivariate function should be selected; a simple choice is the bivariate Poisson function. The non-normalized bivariate Poisson distribution is specified as

$$f(n_1, n_2, \rho) = \mu_1^{n1}\mu_2^{n2}(n_1!n_2!)^{-1}\left[1 + \rho(\mu_1\mu_2)^{1/2}p_{11}p_{21} + \ldots\right], \tag{13.30}$$

where n_1 and n_2 are the degrees of time- and space-lags and where ρ is the correlation coefficient between n_1 and n_2, p_{11} and p_{21} being the Charlier-type B polynomials (Ord 1972) defined as

$$p_k = \sum_{j=0}^{j=k} (k_j)\mu^{k-j}(-1)^j n_j, \tag{13.31}$$

with n_j denoting the frequency corresponding to index j. In the present case, n_{1j} and n_{2j} represent the relevant (space *and* time) lags. This model has been applied to

11 Belgian spatial units over the years 1995–2002; the following results demonstrate its applicability.

For the Belgium empirical example, parameters μ_1 and μ_2 are strictly positive, at the same time showing widely divergent values; ρ lies between –1 and +1, as it is required to do. All spatial parameters are positive, again varying widely in value, as are time parameters, but with three exceptions. A mixed space–time parameter—together with ρ, and typical for the space–time effect—is mostly negative, with four exceptions, the same applying for ρ, with three exceptions. In terms of the lag coefficients, two spatial lags are peaking at contiguity 4, one at contiguity 3, and three at contiguity 2; for all other regions, the impact declines with increasing spatial lag. Overall, this array of values represents a striking variety of effects, due to the specific characteristics of the local economies considered. This is one of the main features of applied spatial econometrics, a point encountered frequently in spatial econometric work. In terms of the time lags, the farthest away peak is 7; another region peaks at 3, and four others at 2. All other regions show a declining time effect. These latter results are classic, but the first ones mentioned are really astonishing. As to ρ of Eq. (13.30) and the mixed space–time parameter, this being the first occasion in which they seem to appear in combined space–time dynamics, at this time, little can be said about their relative values, as more experience with them has to be gathered.

13.2.2.5 Applications

In this section, the discussion in Sect. 13.2.2.4 is applied to the simulation results reported in Sect. 13.2.2.3.

Models (13.20) and (13.21)

A Poisson distribution specification is used as a spatial discount function; recall that its functional form is $f(n) = e^{-\mu}\mu^n/n!$, where n denotes the degree of a time lag.

Again using the data appearing in Table 13.3, and estimating by SDLS—with the weighting mentioned in Sect. 13.2.2.3—results in $a = 0.9524$, $a^* = 1.2058$, and $\mu = 0.0268$, outcomes hardly different from those contained in Models (13.20) and (13.21) of Sect. 13.2.2.3, except for a slight increase in a. But the small value of μ most likely indicates that no *positive* second-order effect exists, which invites trying a negative value for that effect, which, in turn, would lead to the values of the parameters in Eq. (13.15).

Models (13.22) and (13.23)

The quantities appearing in Table 13.4 are employed again, once more with the weighting of Sect. 13.2.2.3. Because only one neighboring spatial unit exists (the other series), no bivariate Poisson distribution can be used. However, two different distributions have been used, one for the own time lag, the other for the time lag of the other series. Table 13.6 presents the SDLS estimation results of the parameters,

Table 13.6 Parameters of re-estimating Eqs. (13.27a) and (13.27b)

Parameters	Series 1	Series 2
a	0.3955	0.1049
b	0.7803	0.7332
c	−0.1632	0.1949
d	−0.1117	0.4073
m	0.0354	0.3931
μ^*	0.5646	1.7152

whose meanings are the following: a (own reaction parameter), b (pure contemporaneous spatial effect), c (reaction parameter to the other series), d (constant), μ (own Poisson lag parameter), and μ^* (Poisson lag parameter of the other series); no correlation coefficient is included because the two Poisson effects are not joint.

Considering the first series, coefficient a is practically equal to that of Eq. (13.27a), 0.40, with a dominating first-order effect ($\mu < 1$). The coefficient for the other region (c) is negative, which is compensated for by the pure contemporaneous spatial effect (b), which has to be taken into account in the dynamics due to the series serial correlation; once more, the Poisson parameter indicates a dominating first-order lag.

As for the second series, the effect of the other series is dominating, simultaneously through b and c, with a significant second-order lag effect ($\mu^* > 1$). Without reproducing exactly Eq. (13.27b), the overall tendency of that equation is confirmed.

13.2.2.6 A General Approach

The problem still can be approached from a more general point of view. Suppose there are observations over R regions and T time periods; excluding a region's own contemporary coefficients—for identification reasons—and also its expectation—forward—effects, $C = R^2T$ coefficients are estimable, including region-specific constants. Because $O = RT$ observations exist, $M = RT(R - 1)$ missing observations remain.

Supposing all lagged observations to be available (these observations do not appear as endogenous variables; see the remark pertaining to Table 13.3), two possibilities remain. Either the parameters are considered to be non-region-specific, in which case $R - 1 = 0$, and hence all coefficients can be interpolated, or RT region-specific parameters may be selected on the basis of an appropriate criterion. In the exercise investigated in this section, the variation coefficient of the absolute parameter values, σ/μ, has been maximized, in order to obtain maximum contrasting impact effects (using real values might imply a μ near zero, with possible indeterminacy; this was the case with the quantities in Table 13.4, which render $\mu = 0.0231$ and $\sigma/\mu = 56.8695$).

Table 13.7 Artificial data for an application

TR	1	2	3
–1	4	2	10
0	5	4	11
1	7	6	13
2	8	6	14
3	10	7	14

Table 13.8 Artificial data calculation results

	Regression			
Coefficient	All *R*	R_1	R_2	R_3
a_1	–0.3110		–0.4091	
a_2	0.8645			1.8333
a_3	–0.3110	–0.1500		0.6667
a_4	0.8645	0.0300		–2.1667
a_5	0.1652			
a_6	–0.1561	0.0700		
a_7	17.1641			
a_8	–1		0.0500	
a_9	–0.5237		0.3182	
σ/μ	–		0.8632	

These two preceding possibilities were applied to the numbers appearing in Table 13.7.

Only time periods 1, 2, and 3 have been used in calculations, periods 0 and –1 only allow estimation of the lag parameters without discounting (see Sect. 13.2.2). Table 13.8 presents the results.

Column *R* in Table 13.8 refers to the first possibility (non-region-specific parameters), whereas columns R_1, R_2, and R_3 refer to the other case in which parameters are region-specific (the nine equations in the system comprise three groups of three separate equations, appropriate binary variables assuring the right split). The parameters a_1 and a_2 are the region's own first- and second-order lag coefficients; a_3 and a_4 apply to the same lags for the effects $R_1 \rightarrow R_2$, $R_2 \rightarrow R_1$, $R_1 \rightarrow R_3$, and a_5 and a_6 to those for $R_3 \rightarrow R_1$, $R_3 \rightarrow R_2$, $R_2 \rightarrow R_3$; a_7 is the constant; a_8 is the spatial autoregression coefficient for $R_2 \rightarrow R_1$, $R_1 \rightarrow R_2$, $R_1 \rightarrow R_3$, and a_9 the one for $R_3 \rightarrow R_1$, $R_3 \rightarrow R_2$, $R_2 \rightarrow R_3$.

The following is a typical equation, where x refers to R_1, y to R_2, and z to R_3:

$$x = a_1x_{t-1} + a_2x_{t-2} + a_3y_{t-1} + a_4y_{t-2} + a_5z_{t-1} + a_6z_{t-2} + a_7 + a_8y_t + a_9z_t \quad (13.32)$$

As for the coefficients, large differences occur according to the specification in question and also the absence of a constant in the region-specific case. This outcome again is an example of multiple regimes ruling interregional behavior

(Griffith and Paelinck 2011, Chap. 13). Of note also is that the region-specific parameters should include the effects of the relevant topology.

13.2.2.7 Conclusions

Only a few tests are available concerning the use of flexible spatiotemporal weight functions (see the previously mentioned Ancot–Paelinck and Griffith–Paelinck studies), but they tend to show that the choice of such functions can correctly specify complex space–time reaction patterns. An obvious extension would be the choice of more complex functional expressions (e.g., bivariate polynomials), a topic for further investigation.

As for the region-specific case treated in Sect. 13.2.2.6, the availability of large T multiregional series should allow an increasing of the number of region-specific parameters, limiting, for example, the lags to only a few orders.

13.3 Chapter Conclusions

As in physics, time, as approached in this chapter, cannot be ignored in economic analysis, and in its spatial approach, space–time combinations creep up naturally.

In this complex matter, misspecifications are to be expected. Hence, the importance of corrective actions and the analysis of corrective issues. Spatial discount functions reveal themselves again as flexible corrective procedures, taking care of regional specificities that should be considered systematically in spatial econometric exercises.

References

Agarwal, V. (2009). *The impact of trade liberalization on income inequality in India*. Saarbrücken: VDM Verlag.

Ancot, J.-P., Chevailler, J., Paelinck, J., Smit, H., & Stijnen, H. (1978). Parameter-component models in spatial econometrics. *The Econometrics of Panel Data, Annales de l'INSEE, 30/31*, 83–98.

Ancot, J.-P., & Paelinck, J. (1983). Interdépendances spatiales et modèles économétriques du développement multirégional. In B. Planque (Ed.), *Le développement décentralisé: dynamique spatiale de l'économie et planification régionale* (pp. 107–116). Paris: LITEC.

Asada, T., & Ishikawa, T. (2007). *Time and space in economics*. Tokyo: Springer.

Balestra, P. (1982). Dynamic misspecification and serial correlation. In J. Paelinck (Ed.), *Qualitative and quantitative mathematical economics. Advanced studies in theoretical and applied econometrics series* (Vol. 1, pp. 115–145). The Hague: Martinus Nijhoff.

Barbour, J. (2000). *The end of time*. New York: Oxford U. Press.

Barker, T. (1996). *Space-time economics*. Cambridge: Cambridge U. Press.

Barreau, H. (2005). *Le temps* (3rd ed.) ed.). Paris: Presses Universitaires de France.

Beguin, H., & Thisse, J.-F. (1979). An axiomatic approach to geographical space. *Geographical Analysis, 11*, 325–341.
Brans, J.-P., Stengers, I., & Vincke, P. (1988). *Temps et Devenir*. Geneva: Patiño.
Casti, J. (1997). *Would-be worlds*. New York: Wiley.
Corpataux, J., & Crevoisier, O. (2007). Economic theories and spatial transformations: Clarifying the space-time premises and outcomes of economic theories. *Journal of Economic Geography, 7*, 285–309.
Couderc, P. (1954). *La Relativité*. Paris: Presses Universitaires de Frances.
de Jong, F. (1967). *Dimensional analysis for economists*. Amsterdam: North-Holland.
Fadden, A. (1977). *Economics of space and time: The measure-theoretic foundations of social science*. Ames, IA: The Iowa State University Press.
Gould, S. (1987). *Time's arrow, time's cycle*. Cambridge, MA: Harvard University Press.
Gray, J. (1979). *Ideas of space*. Clarendon: Totton-under-Edge.
Greene, B. (2004). *The fabric of the cosmos*. New York: Random House.
Griffith, D., & Paelinck, J. (2009). Specifying a joint space- and time-lag using a bivariate Poisson distribution. *Journal of Geographical Systems, 11*, 23–36.
Griffith, D., & Paelinck, J. (2011). *Non-standard spatial statistics and spatial econometrics*. Berlin: Springer.
Hawking, S. (1989). *A brief history of time*. Toronto: Bantam Books.
Hestenes, D. (1966). *Space-time algebra*. New York: Gordon and Breach.
Heyde, J. (1957). *Entwertung der Kausalität?* Stuttgart: Kohlhammer.
Huriot, J., & Perreur, J. (1990). Distances, espaces et représentations, une revue. *Revue d'économie régionale et urbaine, 2*, 197–237.
Kocornik-Mina, A. (2007). *The effects of space on inter-state growth dynamics and income disparities in India: Modeling the simultaneous growth of a system of spatial units*. Unpublished doctoral dissertation, School of Public Policy, George Mason University, Fairfax, VA.
Lesse, P. (1983). Economics in time and space: A mathematical theory. *Papers of the Regional Science Association, 11*, 81–118.
Maddison, A. (1995). *The economic performance of nations. Essays in time and space series*. London: Edward Elgar.
Melvin, J. (1990). Time and space in economic analysis. *The Canadian Journal of Economics, 23*(4), 725–741.
Ord, J. (1972). *Families of income distributions* (Statistical monographs & courses series, No. 30). London: Griffin.
Paelinck, J. (2000). On aggregation in spatial econometric modelling. *Journal of Geographical Systems, 2*(2), 157–165.
Paelinck, J., Ancot, J.-P., & Kuiper, J. (1983). *Formal spatial economic analysis*. Aldershot: Gower.
Paelinck, J., & Mur, J. (2013). Theoretical spatial economics and spatial econometrics: Time-and-space non-convexities galore. *Estadistica Espñola, 55*(180), 13–32.
Prigogine, I., & Stengers, I. (1988). *Entre le temps et l'éternité*. Paris: Fayard.
Pucelle, J. (1955). *Le temps*. Paris: Presses Universitaires de France.
Rovelli, C. (2008). *Temps, Espace*. Paris: Gilson.
Sheffrin, S. (1996). *Rational expectations* (2nd ed.). Cambridge: Cambridge U. Press.
Varii Auctores. (2000). *Les sciences du temps, Lettre 41*. Brussels: FNRS.
von Franz, M.-L. (1978). *Nombre et Temps*. Paris: La Fontaine de Pierre.

Chapter 14
Hybrid Dynamical Systems and Control

Hybrid Dynamical Systems (HDS) have been intensively studied in engineering; some of the specifications developed there could be of use in spatial econometrics. This chapter investigates two such cases gated.

Coutrot et al. (2009) already apply a HDS, a (variously) constrained dynamic system, as a spatial econometric specification. In this chapter, some further applications are investigated: Sect. 14.1 presents a theoretical model, and Sect. 14.2 presents an application to spatial econometrics proper.

Two notions are used later on: a *guard,* which specifies the subset of state space where a certain transition is enabled, and a *reset map*, which is a—possibly—set-valued function, which specifies how new states are related to previous states after a particular transition.

The borrowing from operations research is the extra formal ingredient to this chapter.

14.1 A Theoretical Model

Consider the following system:

$$x = -b, \tag{14.1}$$

$$x \geq 0. \tag{14.2}$$

A physical example is the bouncing ball model. It can be shown (De Schutter and Heemels 2011, p. 22) that events (in this case return drops to earth) occur at discrete times such that

D. A. Griffith, J. H. P. Paelinck, *Morphisms for Quantitative Spatial Analysis*,
Advanced Studies in Theoretical and Applied Econometrics 51,
https://doi.org/10.1007/978-3-319-72553-6_14

$$t_i = t_{i-1} + ab^{-1}c^{i-1}, \tag{14.3}$$

where b is the gravity constant, a is the initial (upward) velocity, and c is the Newton restitution factor (in the bouncing ball case, a function of the ball's elasticity in the physical sense; $0 < c < 1$). The movement stops at a finite time such that

$$t^0 = ab^{-1}(1-c)^{-1}. \tag{14.4}$$

In the spatial economics case, the model could be given the following interpretation (for another example of such a switch, see Barnett 2004): a is an initial stimulus (e.g., the creation of a new activity complex in some spatial unit, with level a), b is the resistance to adoption, and c is the spatial friction factor.

Several cases of surrounding regions can be considered. If the spatial unit is a square, completely surrounded by square units adopting part of the initial impulse, the total effect is

$$e = a\left[1 + 8b^{-1}c(1-c)^{-2}\right]. \tag{14.5a}$$

In the case of a Manhattan circle (see Sect. 2.3.1), this total effect is

$$e = a\left[1 + 4bc^{-1}(1-c)^{-2}\right] \tag{14.5b}$$

and in the case of a hexagonal spatial structure,

$$e = a\left[1 + 6b^{-1}c(1-c)^{-2}\right]. \tag{14.5c}$$

Again, as in the physical case, the process stops in finite time, in the spatial case with a finite total effect.

If the process stops after a finite number of steps less than the value of Eq. (14.4), a correction can be applied; in the Manhattan circle case, the total effect then turns out to be, if only three steps are relevant,

$$e = a\left[1 + 4b^{-1}c(1-c)^{-2}\left(1-4c^3\right)\right]. \tag{14.5b*}$$

14.2 A Spatial Econometric Specification

Consider the following Blondel–Tsitsiklis model (De Schutter en Heemels, 2011, p. 14):

$$x_{i+1} = x_i + A_1x_i \quad \text{if } c'x_i \geq 0, \tag{14.6a}$$

$$= x_i + A_2 x_i \quad \text{if } c' x_i < 0, \tag{14.6b}$$

where Eqs. (14.6a) and (14.6b) are the so-called invariants and the inequalities the so-called guards; no reset map (instructions for specification changes) is present here, because the transition takes place instantaneously. As such, this model is even more flexible than the multiple gap one (Paelinck 2002).

Combining this preceding specification with that of Chap. 11 in Griffith and Paelinck (2011), one can write a spatial econometric generalized Lotka–Volterra model as

$$\Delta' \ln x_t = c_{t-1} A_1 x_{t-1} + (1 - c_{t-1}) A_2 x_{t-1}, \tag{14.7}$$

with $c_t(0 \leq c_t \leq 1)$ a possibly varying quantity playing, at the same time, the role of guard and reset map; $\mathbf{\Delta}'$ is a backward difference operator.

This model has been applied to the four Dutch macro-regions (North, East, West, South: NNL, ENL, WNL, SNL; see Fig. 14.1) with c_t being the national growth rate between consecutive times t and $t + 1$, scaled to 0 – 1. The objective was to analyze how much interregional reactions depended on overall growth, a weighted average of local growth rates, whereas in the inequalities in Eqs. (14.6a) and (14.6b), the **c** vector is constant.

Table 14.1 reproduces the quantities (source: STATLINK) used in the (long-term, 1982–2009) application; A, B, C, and D are GRPs [Gross Regional Products (N, E, W, S from Fig. 14.1)]; F are growth rates of national GDP (Gross Domestic Product). Up to 1993, the quantities are in current prices (no quantities at fixed prices being available), quantities for 1994 are interpolated, and the remaining ones at fixed prices.

Table 14.2 shows OLS results, regressands having been divided by 10^6. No special estimation procedure (e.g., SDLS) has been utilized at this stage. The coefficients downward follow the order of the regions; they portray the rows of the A1, A2 matrices in Eqs. (14.6a) and (14.6b).

Although all of the R^2s are highly significant, the parameter values are not necessarily so, which is understandable given the erratic behavior of the finite differences regressand.

The following remarks are also in order. The own endogenous growth parameter of the regions is variable and changes with overall growth. The autonomous growth parameters are mostly positive, except in three cases. The positive dependence on Region W is characteristic, except in the low c_t-case for Region N, which underscores the leading role of Region W in the Dutch regional economy. Finally, all of the other parameters show variability in function of the spatial unit and the overall growth. Meanwhile, stronger conclusions can only be expected after more econometric investigation, in particular with another estimator; work on this topic is presented hereafter in Table 14.3 (*SDLS* estimation; Griffith and Paelinck 2011, Chap. 11).

Of note is that the order of magnitude of the parameters is not the same as that of the quantities in Table 14.2, because levels rather than growth rates are involved

Fig. 14.1 The four Dutch macro-regions

here; the regional and overall R^2s approach 1. As far as the own dynamics are concerned, for high c_ts, the signs correspond to those of Table 14.2, but differ for low c_ts. The influence of Region W on the other regions has the same sign. Finally, the autonomous growth parameters differ considerably.

Given that the parameters appearing in Table 14.3 are practically obtained by interpolation for 68 degrees of freedom—nearly 30 years of observations—the results of that table should be considered as the most reliable.

Table 14.1 Quantities

Year	A: N GRP	B: E GRP	C: W GRP	D: S GRP	F: GDP growth rates
1983	23,718	28,052	82,422	32,355	****
1984	24,488	28,907	85,583	33,670	0.037575
1985	25,449	29,841	89,224	35,570	0.046276
1986	27,534	31,401	92,295	37,495	0.047772
1987	24,363	33,260	97,753	40,238	0.028532
1988	21,694	34,110	10,0206	41,289	0.006781
1989	21,128	35,408	105,291	43,418	0.047084
1990	22,504	38,060	110,168	46,597	0.048334
1991	24,296	40,810	116,832	49,317	0.063079
1992	26,474	42,659	122,123	51,610	0.049131
1993	27,046	45,133	127,660	54,222	0.042456
1994	27,395	46,709	131,641	55,206	0.026783
1995	27,723	48,150	137,427	56,613	0.028065
1996	28,050	49,591	143,213	58,020	0.027295
1997	29,750	51,122	147,604	60,125	0.034285
1998	31,007	53,711	154,115	64,825	0.050857
1999	31,424	56,968	164,666	68,515	0.057323
2000	32,064	60,711	175,178	73,868	0.061063
2001	35,550	65,688	188,498	80,030	0.078583
2002	38,679	70,195	199,969	84,285	0.061265
2003	39,671	72,865	210,708	87,313	0.04338
2004	41,213	75,270	216,036	89,039	0.026443
2005	41,688	76,784	222,305	91,222	0.024466
2006	44,464	79,590	232,399	95,124	0.044323
2007	47,900	84,565	240,841	99,738	0.046443
2008	49,598	90,184	255,933	105,644	0.058134
2009	55,097	93,623	363,331	108,938	6.638391

Table 14.2 Parameter values

Spatial units → c_t values↓	N	E	W	S
High	24.7485	−2.2751	−1.4316	−9.9958
	−66.2411	−7.9318	6.7941	19.1629
	2.9570	3.7975	−3.3396	4.5803
	38.8180	−1.2381	3.1119	−23.3340
	−60.6595	7.3999	4.3718	19.3681
Low	138.0566	−9.9821	−6.9130	−41.4054
	6.0137	−6.7492	1.0079	−3.2560
	−105.3710	21.2593	0.9971	34.8501
	−0.0354	0.0977	0.0602	0.3226
	0.2304	−0.0333	0.0056	−0.1616
R^2	0.7579	0.9338	0.9499	0.9559

Table 14.3 Results of SDLS computations

Spatial units → c_t values ↓	N	E	W	S
High	0.2127	0.0574	0.0040	0.1205
	0.1120	−0.8495	0.2805	−0.0098
	0.5329	0.0321	−0.5660	0.3123
	−0.8640	0.7897	0.2930	−0.1979
	0.5100	−0.5371	0.1389	−0.0546
Low	−0.7946	−0.0016	0.0791	0.2316
	1.1627	−0.0016	−0.0424	−0.2370
	−0.0515	0.0929	−0.0424	0.0964
	−0.2969	−0.1139	−0.0237	−0.0860
	−0.6304	0.1396	0.3547	−0.0524

14.3 Control

The first starting point is the following potentialized partial differential equation (13.8), studied theoretically in Kaashoek and Paelinck (Kaashoek and Paelinck 1994, 1996, 1998, 2001):

$$f(x,t) = \alpha^2 \int_a^b w(\mathrm{x},\xi) f''(\xi,t) \mathrm{d}\xi, \tag{14.8}$$

from which an optimal control for minimal cost can be derived as (see Feichtinger 1992; Kaashoek and Paelinck 1998)

$$u^0(x,t) = -\beta e^{\gamma t} \int_a^b w(x,\xi) \left[\pi_1(x) e^{\varepsilon(x)t} + \pi_2(x) e^{\eta(x)}\right] d\xi, \tag{14.9}$$

the controls being combinations of exponential functions of time. However, this approach—although enriching the analysis by a spatial element, the spatial potentializing function—calls for a certain number of remarks (also see Gevaerts 2014), some of which are as follows:

(a) Equation (14.9) reveals that controlling over space implies controlling everywhere; intuitively bizarre is that controlling far-away is a necessity to guarantee effective results. Sometimes, the possible control could be spatially counterintuitive (a classic is regional policy benefitting developed regions more than the less developed ones that should be favored). International trade is an early example of these interdependencies; see Pöyhönen (1963).
(b) In practice, the spatial units are discrete, so discrete models are to be preferred in applications (see an example in Kaashoek and Paelinck 1998, pp. 18–19).
(c) In the cases commented on in remarks *a* and *b*, only one control is present [the same applies to HDS, because processes are controlled successively; see De Schutter and Heemels (2011)]; again, in practice, multiple and interdependent

objectives have to be controlled simultaneously, so the underlying model has to be extended to comprehend all of the relevant variables, and grand classics apply again (Tinbergen 1956; Theil 1956).

(d) In practice, once more, controls more often than not are operated by independent units (e.g., cities, districts, states); thus, optimality only can be obtained by consensus.
(e) Finally, in practice yet again—such as in the case of HDS—controls are to be instantaneously
(f) tuned, as a function of deviant behavior of parts of the given system.

This context leads to an entirely different model specification

$$G(x, u) = 0, \tag{14.10}$$

$$\varphi(\mathbf{x}). \tag{14.11}$$

Equation (14.10) specifies the functioning of the economies, $\mathbf{x}$ being the relevant variables (possibly over time) and $\mathbf{u}$ the controlling variables. Function (14.11) is a preference function to be optimized over $\mathbf{x}$, functions (14.9) then defining the optimal controls.

If point *d* of the preceding remarks applies, then function (14.11) has to be substituted by

$$\Psi(\mathbf{x}), \tag{14.12}$$

which has to be implemented by, for example, a negotiation process; the next section presents more about this situation.

14.4 Negotiation

Hypergraph Conflict Analysis (HCA) was introduced by Paelinck and Vossen (1983). Since then, a certain number of further contributions have appeared, in particular three studies about hypergraph conflict resolution (or reduction: de Koster and Paelinck 1985; van Gastel and Paelinck 1992; Paelinck 2015).

Suppose there are a number of agents, or groups of agents (A_i), confronted with a set of possible options (O_j), whatever the latter may be. In spatial analysis, one might encounter infrastructural projects, regional development issues, and many other strategic choice problems. The agents could agree or disagree with some of the options, and this state of affairs can be set out in a table or matrix (**C**), with Table 14.4 reproducing a 3-by-3 binary (only full agreement or disagreement is assumed here) case.

The hypergraph nature of Table 14.4 results from the fact that for each agent, the agreeable options are a subset of the overall set of possible options (also see Voloshin 2009).

Table 14.4 Matrix C

Agents\options	O_1	O_2	O_3
A_1	1	0	0
A_2	0	1	0
A_3	0	0	1

Several measures can be proposed to show the degree of conflict. One, denoted here by δ, divides the minimum number of zeroes taken over the columns of matrix **C** by the number of agents, n_i (for Table 14.4, this is 2/3). Another measure, denoted by τ, is the transversal number, defined as the cardinal of the minimal set of options on which all agents taken together agree (for Table 14.4, this number is 3). A relative transversal number, τ^*, would divide $\tau - 1$ by the cardinal of the set of potential options minus one (for Table 14.4, $\tau^* = 1$). Clearly, that δ and τ (or τ^*) are interrelated.

The transversal number can be computed by solving the following program:

$$\min I'o, \tag{14.13}$$

$$Co \geq i \tag{14.14}$$

$$o = \widehat{o}o, \tag{14.15}$$

where **i** is the unit column-vector, **o** is a column-vector of binary variables (conditions 14.13), $\hat{\mathbf{o}}$ is its diagonal matrix, and matrix **C** contains the elements of the negotiation hypergraph.

If $\delta = 0$, or alternatively, $\tau = 1$ (or $\tau^* = 0$), no conflict exists, because all agents agree on at least one option. Hypergraph conflict resolution aims at computing an optimal way of turning over agents, so as to drive δ down to zero or τ to 1. The following development shows how this can be done for a reduction of τ with one unit.

Defining an agent's resistance coefficient (resistance to changing initial positions) over its cells as $r_{ij,}$ forming a column-vector $\boldsymbol{r}$; a total resistance function, to be minimized, may be written as

$$\rho = r'c, \tag{14.16}$$

where $c =$ **vecC,** $c_{ij} = 1$ implying $r_{ij} = o$. The following constraints apply:

$$C\widehat{o}i \geq i, \tag{14.17}$$

$$\mathbf{i}'\mathbf{o} = \tau - 1, \tag{14.18}$$

$$o = \widehat{o}o, \tag{14.19}$$

the variables being defined as for equation system (14.13)–(14.15), condition (14.17) stating that every agent should agree with at least one option in the new set of $\tau - 1$ options.

An obvious result for Table 14.4 is $c_{ij} = 1$ for min r_{ij}, which reduces τ to 2 (e.g., options 1 and 2). For larger negotiation matrices (more agents and/or options), the end result would be less obvious. From here-onwards, the step can be repeated to bring τ down to 1. Corresponding policies (controls) then can be computed according to Eq. (14.9). A topographical representation of the preceding results can be found in Paelinck and Mur (2013).

Fuzzy generalization of these ideas is immediate, but this topic is not addressed here.

14.5 Conclusions

Efforts with HDSs seem promising, because they have been shown before; work along those lines will continue in future research.

One topic to be investigated is the possibility of estimating separately the coefficients *a, b* and *c* of the bouncing ball model of Sect. 14.1 (i.e., initial impulse, resistance to acceptance, spatial friction), possibly the cutting-off point. The other concern is about more systematically introducing in spatial econometric models elements of macroeconomic situations, as tried in Sect. 14.2.

The theoretical result of the analysis presented in this chapter is that there exists something that could be called a negotiable chaos. Moreover, the tools presented in this chapter are sufficiently developed now to be applied to empirical cases; perhaps more important is the focus on spatial patterns, some of them unexpected. The latter feature renders political decisions more intricate to make, because they might have to apply to nonconventional spaces.

References

Barnett, V. (2004). Changing an interpretation: Slutsky's random cycles revisited. *European Journal of the History of Economic Thought, 1*, 5–11.

Coutrot, B., Sallez, A., Paelinck, J., & Sutter, R. (2009). On potentialized partial finite difference equations: Analyzing the complexity of knowledge-based spatial economic activities. *Région et Développement, 29*, 237–264.

de Koster, M., & Paelinck, J. (1985). Reduction of conflict. *Organizational Conflict and Peace Science, 4*, 1–17.

De Schutter, B., & Heemels, M. (2011). *Modeling and control of hybrid systems. Lecture notes for the DISC course*. Delft: Delft University of Technology. Accessed October 15, 2017, from http://disc.tudelft.nl/wp-content/uploads/2016/07/2017-modeling-and-control-of-hybrid-systems.pdf

Feichtinger, G. (1992). *Dynamic economic models and optimal control*. Amsterdam: Elsevier.

Gevaerts, C. (2014). *Formal abstraction of stochastic systems*. Delft: Centre for Systems and Control, Delft University of Technology.

Griffith, D., & Paelinck, J. (2011). *Non-standard spatial statistics and spatial econometrics*. Berlin: Springer.

Kaashoek, J., & Paelinck, J. (1994). On potentialised partial differential equations in theoretical spatial economics. In D. Dendrinos (guest ed.), Special issue of *Chaos, Solitons and Fractals* entitled "Non-Linear Dynamics in Urban and Transport Analysis," *4*(4), 585–594.

Kaashoek, J., & Paelinck, J. (1996). Studying the dynamics of pre-geographical space by means of space and time-potentialised partial differential equations. *Geographical Systems, 3*, 259–277.

Kaashoek, J., & Paelinck, J. (1998). Potentialised partial differential equations in economic geography and spatial economics: Multiple dimensions and control. *Acta Mathematica Applicanda, 51*, 1–23.

Kaashoek, J., & Paelinck, J. (2001). Potentialised partial differential equations in spatial economics: Some further results on the potentialising function. *The Annals of Regional Science, 35*(3), 463–482.

Paelinck, J. (2002). A multiple gap approach to spatial economics. *The Annals of Regional Science, 36*(2), 219–228.

Paelinck, J. (2015). Representing negotiation space. In P. Nijkamp, A. Rose, & K. Kourtit (Eds.), *Regional science matters: Studies dedicated to Walter Isard* (pp. 273–286). Cham: Springer.

Paelinck, J., & Mur, J. (2013). Theoretical spatial economics and spatial econometrics: Time-and-space non-convexities galore. *Estadistica Espñola, 55*(180), 13–32.

Paelinck, J., & Vossen, P. (1983). Axiomatics of conflict analysis. In W. Isard & Y. Nagao (Eds.), *International and Regional Conflict* (pp. 33–52). New York: Ballinger.

Pöyhönen, P. (1963). A tentative model for the volume of trade between countries. *Weltwirtschaftliches Archiv, 90*, 93–99.

Theil, H. (1956). *Optimal decision rules for business and government*. Amsterdam: North Holland.

Tinbergen, J. (1956). *Economic policy: Principles and design*. Amsterdam: North-Holland.

van Gastel, M., & Paelinck, J. (1992). Computing the solution to a conflict situation by means of continuous multicriteria analysis. In M. Chatterji & L. Forcey (Eds.), *Disarmament, economic conversion and the management of peace* (pp. 27–39). New York: Praeger.

Voloshin, V. (2009). *Introduction to graph and hypergraph theory*. Hauppang, NY: Nova Science.

Chapter 15
The W Matrix Revisited

This chapter summarizes a study of two examples in which it is shown that bridging some specifications—belonging to different fields of investigation—can lead to simplifying and at the same time enriching, a spatial model structure; that bridging is the formal contribution of this chapter.

The next section, Sect. 15.1, previously was published in *Estadistica Española*, the official journal of the Spanish Statistical Institute, and reproduced here with kind permission.

15.1 Consistent Spatial Modeling

The topological structure of the set of spatial units can assist in identifying certain model parameters; this section shows how this idea can be applied to a complete spatial econometric model.

Consider the extended spatial simultaneous autoregressive (SAR) model, which may be written, using matrix notation, as

$$\mathbf{y} = \mathbf{A}\mathbf{y} + \mathbf{A}^{*}\mathbf{X}\mathbf{b} + \boldsymbol{\varepsilon}. \tag{15.1}$$

One of its noteworthy features is that the **y** variable values all have the same definition, GRP, for instance, contrary to the classical nonspatial model $\mathbf{A}\mathbf{y} + \mathbf{X}\mathbf{b} = \boldsymbol{\varepsilon}$, where the vector **y** may well consist of different variables. Its second noteworthy feature is that model (15.1) is isomorphic to the classical input–output model:

$$\mathbf{y} = \mathbf{A}\mathbf{y} + \mathbf{f}, \tag{15.2}$$

where matrix **A** is the input coefficients matrix, and **f** is the final demand vector.

The difficulty with the spatial econometric model, contrary to the input–output one, is that the matrix **A** coefficients of equation (15.1) are unknown *a priori*; but

D. A. Griffith, J. H. P. Paelinck, *Morphisms for Quantitative Spatial Analysis*, Advanced Studies in Theoretical and Applied Econometrics 51,
https://doi.org/10.1007/978-3-319-72553-6_15

from Eq. (15.2), total relative inputs can be computed as $\mathbf{a}' = \mathbf{i}'\mathbf{A}$, which leads to the following suggestion for the spatial econometric model.

The assumption is that total *received* impacts by a spatial unit i (a_i from the preceding vector $\mathbf{a}'$) are a function of its production level; the higher that level, the higher the impacts. This specification allows for cross influences of receiving and impacting regions, the latter resulting from the first right-hand side of Eq. (15.1).

The proposed function is

$$a_i = 1 - e^{-\alpha y(i)}, \tag{15.3}$$

where y_i is the production level of region i; of note is that the exponent in Eq. (15.3) compensates dimensionally. The coefficient α is to be estimated jointly with the coefficients of matrix $\mathbf{A}$.

Next model (15.1) is applied to the contiguity data of Table 15.1 (which furnishes a starting point for a matrix $\mathbf{W}$) and GRP data (10^5 in 2000 Euros) of Table 15.2, both for Belgium.

The spatial units have then been split into three categories: high (A, BC), medium (VB, OV, WV, LIM, H, LIE), and low (BW, N, LU) GRPs, with 11 a_{ri} coefficients: a_1, a_2, a_3: zero order contiguity (own) impact; a_4, a_5, a_6: first order; a_7, a_8, a_9: second order; and, a_{10}, a_{11}, a_{12}: third order. The so created mixed contiguity order has been translated into the matrix $\mathbf{A}$, allowing the 11 constraints (15.3) to be set.

Table 15.1 Contiguity degrees of the Belgian provinces

Provinces	A	BW	VB	OV	WV	LIM	H	N	LU	LIE	BC
A	0	2	1	1	2	1	2	3	3	2	2
BW	2	0	1	2	2	2	1	1	2	1	2
VB	1	1	0	1	2	1	1	2	2	1	1
OV	1	2	1	0	1	2	1	2	3	2	2
WV	2	2	2	1	0	3	1	2	3	3	3
LIM	1	2	1	2	3	0	2	2	2	1	2
H	2	1	1	1	1	2	0	1	2	2	2
N	3	1	2	2	2	2	1	0	1	1	3
LU	3	2	2	3	3	2	2	1	0	1	3
LIE	2	1	1	2	3	1	2	1	1	0	2
BC	2	2	1	2	3	2	2	3	3	2	0

Note: *A* Antwerpen, *BW* Brabant Wallon, *BC* Brussels Capital, *H* Hainaut, *LIM* Limburg, *LIE* Liège, *LU* Luxembourg, *N* Namur, *OV* Oost-Vlaanderen, *VB* Vlaams-Brabant, *WV* West-Vlaanderen

Table 15.2 Gross regional products of the Belgian spatial units, 1995

Units	A	BW	VB	OV	WV	LIM	H	N	LU	LIE	BC
Values	416,028	62,919	211,584	255,118	226,222	143,460	191,433	64,851	37,976	173,063	424,381

Table 15.3 Values of Eq. (15.3)

Units	A	BW	VB	OV	WV	LIM	H	N	LU	LIE	BC
Value	0.926	0.325	0.734	0.797	0.757	0.592	0.698	0.333	0.211	0.661	0.929

Table 15.4 a_{ir} coefficients

Coefficient	a_1	a_2	a_3	a_4	a_5	a_6
Value	0.0954	0.0886	0.0640	0.0837	0.0699	0.0651
Coefficient	a_7	a_8	a_9	a_{10}	a_{11}	a_{12}
Value	0.0790	0.0520	0	0.0921	0.0802	0.0042

Table 15.5 The **X** matrix

Units\sector	Agriculture	Industry	Services
A	0	1	1
BW	1	0	0
VB	1	1	1
OV	1	1	0
WV	1	0	1
LIM	1	0	0
H	0	1	0
N	1	0	1
LU	1	0	0
LIE	0	1	0
BC	0	1	1

Note: See Table 15.1 for unit definitions

Because the system is underdetermined, and moreover might be partially inconsistent, the estimation solution was sought for values guaranteeing consistency, along the following lines: expression (15.2) was minimized (the method so chosen discarding possible outliers), which changed the coefficient of OV to 0.7080, the one for LIM to 0.6901, and the one for LIE to 0.7259; this adjustment, under conditions (15.3), lead to the coefficients appearing in Tables 15.3 and 15.4, the value of coefficient α being 6.2550E−6.

The impact of exogenous variables could then be computed from $\mathbf{y}^* = (\mathbf{I} - \mathbf{A})\mathbf{y}$; **X** from Eq. (15.1) was constructed as a binary three sector (agriculture, industry, services) matrix (Table 15.5).

The result (with OLS) is vector $\boldsymbol{b}$ = [−73161; 109,175; 106,678; constant = −11,687]′, with $R^2 = 0.6807$, significant at the 5% level. The coefficients have admissible signs, agriculture depressing regional products; given the binary construction of matrix **X**, significant coefficients were hardly to be expected.

Now, a very curious may be observed: the correlation between $\boldsymbol{y}$ of Eq. (15.1), and $\mathbf{y}^*$ as defined in the preceding discussion, is extremely high ($R^2 = 0.9968$), and the regression of $\boldsymbol{y}$ on $\boldsymbol{y}^*$ gives a slope of 0.9939 and a constant of 151,123. This outcome means that if one takes vector ***y*** rather than ***y**** as the regressand, then the difference is approximately only a constant. Indeed, one has

$$y = y^* + ci = Xb + ci. \tag{15.4}$$

The regression of **y** on **X** results, as expected, in a slightly higher R^2, and the parameters (−67,785; 115,417; 108,489; *constant* = 131,475) also show slopes of the same order of magnitude as the preceding ones, but with a significant higher constant. This outcome means that the results of Thompson and Mattila (1959) might have been the reduced form of Eq. (15.1), so unwittingly taking into account spatial interdependences.

But what about forecasting? This should be performed in an iterative way, given that matrix **A** depends on vector **y** via Eq. (15.4).

In conclusion, a common rule of certain econometric models—to wit input–output ones—can be an important help—together with topological information (also see Paelinck and Klaassen, 1979, in particular pp. 8–9 and footnote 21)—in identifying and estimating a multiregional model.

15.2 Lotka–Volterra Systems as Generalized Logistic Models

Let us first introduce generalized Lotka–Volterra systems (GLVSs); a generalized GLVS can be written in matrix-vector notation as

$$\dot{\mathbf{u}} \triangleq \widehat{\mathbf{u}}(Au + a), \tag{15.5}$$

where **u** is a column vector of (endogenous) variables, $\widehat{\mathbf{u}}$ its diagonal matrix version, **A** is a square matrix, and **a** is a column vector of fixed coefficients; the • notation denotes the time derivative, ∂/∂t.

Given Eq. (15.5), the variables *u* describe a time path that can take on all of the characteristics of general continuous dynamic processes (e.g., convergence, divergence, and limit circles; see Braun, 1975, Sect. 4.9; Gandolfo 1996, in particular Sect. 24.4; Peschel and Mendel, 1986). A sufficient condition can be derived (Griffith and Paelinck 2011, Chap. 12) for Eq. (15.5) to converge to its focus, the product—$\mathbf{A}^{-1}\mathbf{a}$, by constructing a Lyapunov function (Hahn 1963).

From the differential specification of the classical logistic equation

$$u = a^* u^- (1 - \lambda u), \tag{15.6}$$

one can derive its generalization

$$\dot{\mathbf{u}} \widehat{=} \widehat{\mathbf{u}} A^* \left(i - \widehat{\boldsymbol{\lambda}} y\right). \tag{15.7}$$

In expression (15.6), λ^{-1} is the value of the asymptote, generalized to the diagonal matrix of λs in Eq. (15.7).

The equivalence of Eqs. (15.6) and (15.7) can be shown as follows:
define

$$\mathbf{A}^{*} i \triangleq \mathbf{a} \tag{15.8a}$$

and

$$\mathbf{A}\widehat{\boldsymbol{\lambda}} = A^{*}, \tag{15.8b}$$

i being the unit column vector, these definitions imply

$$\mathbf{y} = \mathbf{A}^{*} \left(\mathbf{i} + \lambda^{-1} \mathbf{y}\right), \tag{15.9}$$

-$\boldsymbol{\lambda}$ being the asymptotic vector. This vector indeed is equal to the classical equilibrium vector (if the system converges) -$\mathbf{A}^{-1}\mathbf{a}$ because

$$\mathbf{a} = \mathbf{A}^{*}\mathbf{i} = \mathbf{A}\boldsymbol{\lambda} \tag{15.10a}$$

and hence

$$-\mathbf{A}^{-1}\mathbf{a} = -\mathbf{A}^{-1}\mathbf{A}\lambda = -\lambda, \tag{15.10b}$$

which implies that no matrix inversion is necessary to compute the possible focal point.

Application of these developments to a four-variable case borrowed from Griffith and Paelinck (2011, Chap. 16) is summarized next. Table 15.6 presents the data. The case is about a two-region two-sector model.

Application of Eqs. (15.1) and (15.1) leads to the vector $\boldsymbol{\lambda}$ = (0.5098, 0.5214, 0.5023, 0.5026); the quantities are shares of sectoral data per spatial unit, such that the possible equilibrium values are fifty-fifty deals. Of note is that the quantities have originally been computed for an integral logistic specification. With regard to convergence, $tr\mathbf{A} = -4.6501$, so that sufficiency conditions for convergence could be satisfied (they are that the real parts of all of the eigenvalues should be strictly negative), but these conditions are only sufficient.

One more interesting point is that Eq. (15.7) can be easily generalized to higher order polynomials such as the following quadratic specification between parentheses (see von Bertalanffy 1968, pp. 60 a.f.):

Table 15.6 Logistic model data

Region combinations\parameters	1	2	3	4	Constant
1,1	−6.1627	44.1253	−6.6102	2.0707	3.3528
1,2	11.7884	3.3036	2.5080	−3.8460	−6.9309
2,1	9.9394	4.1536	−7.9713	−10.9558	2.2372
2,2	−13.3512	-8.3467	3.4470	6.1802	6.0663

Fig. 15.1 A phase diagram

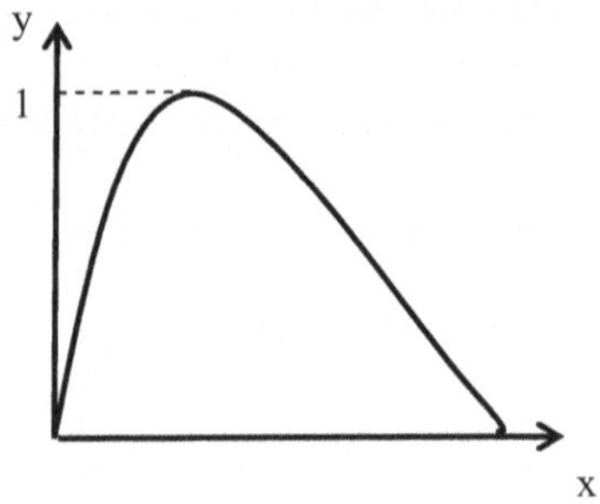

$$\dot{y} = \widehat{\mathbf{y}}A^*(i + \widehat{\boldsymbol{\lambda}}y - \widehat{\boldsymbol{\mu}}\widehat{\mathbf{y}}y). \tag{15.11}$$

The change of signs has been introduced to have an economically meaningful (nonnegative) range for the vector $\mathbf{y}$, which is the range $0,\ y_i^* = \left[0, \mu_i/2\lambda_i\left(1 + \sqrt{1+} \right.\right.$ $\left.\left. 4\mu_i/\lambda_i^2\right)\right]$. Figure 15.1 shows the phase accompanying diagram.

In conclusion, a study of the family links between spatial econometric specifications is a valuable topic to be envisaged for further study. In doing so, an intrusion into general system theory also should be advocated.

15.3 Characterizing the A Matrix in an Extended SAR Model

Recalling the extended SAR model

$$\mathbf{y} = \mathbf{A}\mathbf{y} + \mathbf{A}^*\mathbf{X}\mathbf{b} + \boldsymbol{\varepsilon}, \tag{15.12}$$

some indicators from theoretical physics could be applied to characterize the $\mathbf{A}$ matrix, possibly after adaptation. Consider the vector function

$$\mathbf{v} = [f(\mathbf{u}), g(\mathbf{u}), h(\mathbf{u})], \tag{15.13}$$

with

$$\mathbf{u} = [x, y, z], \tag{15.14}$$

the typical case from theoretical physics being the three-dimensional one. Two notions have been defined (see Johnson and Kiokemeister 1964, pp. 664–666) with regard to this specification. The first is the divergence, defined as

$$\text{div } \mathbf{v} = f'_x + g'_y + h'_z \tag{15.15}$$

and the second is the curl defined as

$$\text{curl } \mathbf{v} = \left[h'_y - g'_z; f'_z - h'_x; g'_x - f'_y\right]. \tag{15.16}$$

The divergence, a scalar, measures the total inflow of a substance into a system, whereas the curl, a vector, measures the net directional flow for a given point (x^*,y^*, z^*). Figure 15.2 portrays this, the starting point for div and curl being point A in that diagram.

Returning to matrix **A** in Eq. (15.12), its diagonal can be nonzero, so that the trace could be interpreted as the divergence. In contrast, as the order of matrix **A,** n (**A**) > 3, the curl could define the differences between the sums of the off-diagonal elements of the minor-cofactor matrices of the diagonal elements.

A more appropriate solution could be the following one; characterize matrix **A** by:

- Its diagonal elements, the separate inflows by coordinate (here the spatial units).
- The difference between the sum of the row corresponding to a diagonal element (the influence from other spatial units) and that of the corresponding column (the outflow to other spatial units), the diagonal element being automatically cancelled; this quantity would measure the net influence of a given spatial unit.

This idea has been applied to an 11-by-11 **A** matrix of 1995 data for Belgium, tabulated as (Table 15.7) here.

Table 15.8 lists the own influence coefficients (divergence, d) and the net impulses (curl, c).

The divergence systematically has the highest coefficient; as to the curls, they are negative for spatial units A, BW, OV, LIM, H, LIE, and BC; four of them (A, H, LIE, and BC) are industrial regions that impact the remaining ones.

Fig. 15.2 Divergence (d) and curl (c)

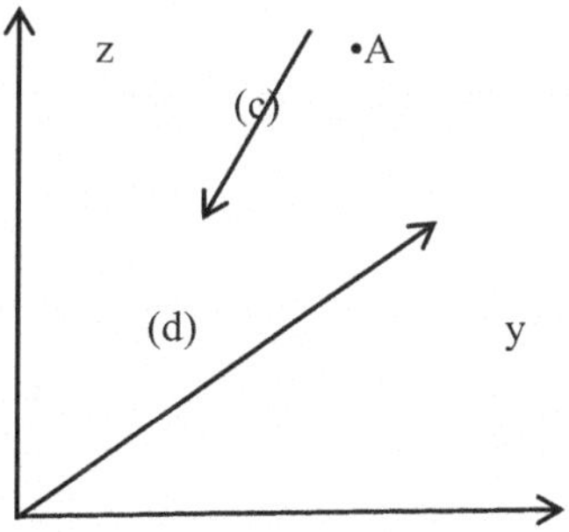

Table 15.7 A matrix of 1995 data for Belgium

Provinces	A	BW	VB	OV	WV	LIM	H	N	LU	LIE	BC
A	0.2334	0.0753	0.0753	0.0753	0.0753	0.0753	0.0753	0.0753	0.0753	0.0753	0.0379
BW	0.0379	0.2474	0.0753	0.0753	0.0753	0.0753	0.0753	0.0753	0.0753	0.0753	0.0379
VB	0.0932	0.0753	0.2474	0.0753	0.0753	0.0753	0.0753	0.0753	0.0753	0.0753	0.0932
OV	0.0932	0.0753	0.0753	0.2474	0.0753	0.0753	0.0753	0.0753	0.0753	0.0753	0.0379
WV	0.0379	0.0753	0.0753	0.0753	0.2474	0.0753	0.0753	0.0753	0.0753	0.0753	0.1486
LIM	0.0932	0.0752	0.0752	0.0752	0.0752	0.2474	0.0753	0.0753	0.0753	0.0753	0.0379
H	0.0379	0.0753	0.0753	0.0753	0.0753	0.0753	0.2474	0.0753	0.0753	0.0753	0.0379
N	0.1486	0.0753	0.0753	0.0753	0.0753	0.0753	0.0753	0.2474	0.0753	0.0753	0.1486
LU	0.1486	0.0753	0.0753	0.0753	0.0753	0.0753	0.0753	0.0753	0.2474	0.0753	0.1486
LIE	0.0379	0.0753	0.0753	0.0753	0.0753	0.0753	0.0753	0.0753	0.0753	0.2474	0.0379
BC	0.0379	0.0753	0.0753	0.0753	0.0753	0.0753	0.0753	0.0753	0.0753	0.0753	0.2334

Table 15.8 Impulses extracted from Table 15.7

Spatial units	Divergence	Curl
A	0.2334	−0.0507
BW	0.2474	−0.0747
VB	0.2474	0.0359
OV	0.2474	−0.0194
WV	0.2474	0.036
LIM	0.2474	−0.0199
H	0.2474	−0.0748
N	0.2474	0.1466
LU	0.2474	0.1466
LIE	0.2474	−0.0748
BC	0.2334	−0.0508

15.4 Conclusions

In spatial econometric analysis, the freely used spatial weights matrix **W** should be given an operational content, in the sense that some underlying economic mechanisms should be present in its (endogenous) derivation. A possibility has been studied in this chapter and use has been made of the findings of another approach, to wit input–output analysis. The application summarized here gives some curious insights into the meanings of partial and total effects of external shocks.

References

Braun, M. (1975). *Differential equations and their applications*. New York: Springer.
Gandolfo, G. (1996). *Economic dynamics* (3rd) ed.). Berlin: Springer.
Griffith, D., & Paelinck, J. (2011). *Non-standard spatial statistics and spatial econometrics*. Berlin: Springer.
Hahn, W. (1963). *Theory and applications of Lyapunov's direct method*. Englewood Cliffs, NJ: Prentice Hall.
Johnson, R., & Kiokemeister, F. (1964). *Calculus with analytic geometry*. Boston, MA: Allyn and Bacon.
Paelinck, J., & Klaassen, L. (1979). *Spatial econometrics*. Farnborough: Saxon House.
Peschel, M., & Mendel, W. (1986). *The predator-prey model*. New York: Springer.
Thompson, W., & Mattila, J. (1959). *An econometric model of postwar state industrial development*. Detroit: Wayne State University Press.
von Bertalanffy, L. (1968). *General system theory*. Harmondsworth: Penguin.

Chapter 16
Clustering: Some Nonstandard Approaches

In this chapter, first a certain number of rigorous definitions are given, this being all the more necessary as the literature abounds in contradictory definitions and sloppy use of terms; for more details about this contention, one is referred to Paelinck (2004). The second part of this chapter describes some relevant clustering methods, including a number of recent analytical developments. For links with transport problems, again consult Paelinck (2004).

The nonstandard treatment in this chapter contributes selected formal additions to the analysis literature.

16.1 An Axiomatic Basis

Axioms are statements assumed to be true that form the foundation for deriving theorems about the context to which they apply. The context of interest here concerns spatial economic landscapes.

16.1.1 Clusters

Assume a set S of arbitrary objects or elements; this set is structured by a distance relation (for a definition, see Paelinck and Vossen 1983, pp. 39–52), such that it is a metric space. A subset of S, S^* may be called a cluster if

$$\forall a \in S^* \exists b \in \; S^* \mid d(a,b) \leq d^*, \tag{16.1}$$

where d is any distance measure. In other words, expression (16.1) means that in order to identify a cluster, one needs to select an S^* from S such that every point in it

D. A. Griffith, J. H. P. Paelinck, *Morphisms for Quantitative Spatial Analysis*,
Advanced Studies in Theoretical and Applied Econometrics 51,
https://doi.org/10.1007/978-3-319-72553-6_16

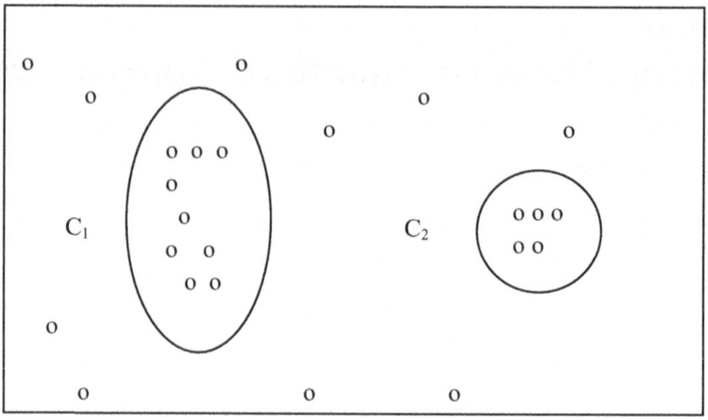

Fig. 16.1 Example clusters

has at least one neighbor at or within distance d^*. The following points are immediate consequences:

- A set S may contain several clusters.
- Given d^*, the clusters present in S do not necessarily exhaust it.
- Decreasing d^* does not necessarily decrease the number of clusters.
- Increasing d^* does not necessarily increase the number if clusters.
- Decreasing d^* does not necessarily decrease the density of the clusters, measured as the number of points over the surface of the convex hull of the cluster points.

Figure 16.1 furnishes a graphical example of this definition; assume that the distance used in it is Euclidean. Increasing d^* can, at a given moment, make clusters C_1 and C_2 join and even lead to some outlying points being integrated into a newly created larger cluster, larger in terms of the number of points present. Real life examples are numerous: urban agglomerations, industrial parks, woods and forests, herds, but also multiregional economic units, administrative entities, and so on.

16.1.2 Complexes

Assume now another type of relation, ρ say, between the elements of a cluster. Accordingly, a subset of a cluster C is called an n-complex if

$$a \;\rho\; b \;\rho\; c \;\rho \ldots \rho\; a, \tag{16.2}$$

i.e., if there exists a closed chain of relations between n elements belonging to the cluster. In still other mathematical terms, the relevant relational matrix, $\boldsymbol{R}$, is indecomposable.

Properties of complexes include the following:

- A threshold put on the intensity of a relation never increases the number of elements belonging to a complex.
- One or more relations can be critical to the size of a complex, because breaking up one relational element can make the complex split into several parts.
- Conversely, introducing a new relationship might dramatically increase the size of the resulting complex; for example, the emergence of a new technological link between two activities disconnected before and linking up previously disconnected clusters.

In other words, these properties mean that if an external impact is exercised on one (or more) of the elements of a complex, its effects will be transmitted to all the elements belonging to that complex. An example of such relations is the well-known input–output linkages (technical links between production units); but other relations (e.g., knowledge transfers, organizational setups) also are instances of this.

16.1.3 Corps

Inside complexes one can possibly isolate several corps, a corps being defined as follows: suppose a measure of the intensity of the previously mentioned relations exists when observing complexes, then a corps is a closed subset of the complex such that

$$\sum\nolimits_{ij}\rho_{ij} = \nu, \tag{16.3}$$

where the ρ_{ij} are the intensity measures of the relations, and ν is their aggregate value. The ensuing Sect. 16.2 treats how these corps are aggregated within a given complex; it is devoted i.e. to linear assignment.

Properties of corps include:

- At a given ν-level, multiple circular (i.e., with their elements linked in a circular chain) corps may be present, without exhausting all of the elements of the complex to which they belong.
- At different ν-levels, different corps may be generated between the elements of the complex considered.

Examples of corps are agro-industrial groups, chemical groups, and metalworking sub-complexes; such groupings have been generated by empirical analysis and are described hereafter.

16.1.4 Hierarchies

Recalling the Tinbergen hierarchies of Chap. 2, more formally, a set H is a hierarchy if:

- It is a set of subsets of a set S.
- The set S belongs to H.
- Each element x_i of S belongs to H.
- For all subsets $A \subset H$ and $B \subset H$, if $A \cap B \neq 0$, then either $A \subset B$ or $B \subset A$.

A hierarchy is called *indexed* if E is a mapping $h(.)$ of H onto [0, 1], satisfying the conditions:

- If A is a single element of S, then $h(A) = 1$.
- if $A \subset B$, then $h(A) > h(B)$.

An example of such a hierarchy is the following:

$$h = 1 \rightarrow \{x_1, x_2, x_3, x_4, x_5\}$$
$$h = 0.8 \rightarrow \left\{\left(x_1, x_2\right\}; x_3, x_4, x_5\right\}$$
$$h = 0.6 \rightarrow \left\{\left(x_1, x_2\right); \left(x_3, x_4\right\}; x_5\right\}$$
$$h = 0.5 \rightarrow \left\{\left(x_1, x_2, x_3, x_4\right); x_5\right\}$$
$$h = 0.2 \rightarrow \{(x_1, x_2, x_3, x_4, x_5)\}$$

To this example corresponds a so-called similarity matrix (e.g., Table 16.1). From Table 16.1, the indexed hierarchy again can be derived. For some (symmetrical) similarity tables, similarity values have to be jumped in order to generate a hierarchy, as is the case for Table 16.2. Figure 16.2 portrays the implicit hierarchy,

Table 16.1 Similarity matrix

Elements	1	2	3	4	5
1	1	0.8	0.5	0.5	0.2
2	0.8	1	0.5	0.5	0.2
3	0.5	0.5	1	0.6	0.2
4	0.5	0.5	0.6	1	0.2
5	0.2	0.2	0.2	0.2	1

Table 16.2 An additional similarity matrix

Elements	1	2	3	4	5
1	1	0.5	0.6	0.2	0.8
2	0.5	1	0.8	0.3	0.7
3	0.6	0.8.	1	0.4	0,6
4	0.2	0.3	0.4	1	0.5
5	0.8	0.7	0.6	0.5	1

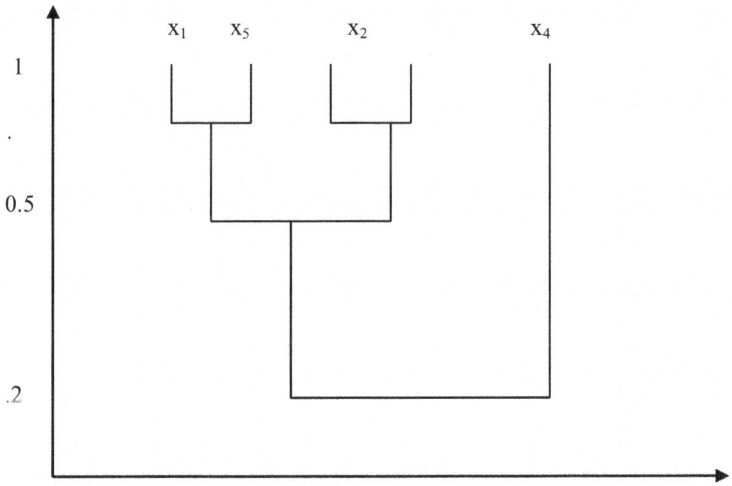

Fig. 16.2 An indexed hierarchy dendogram graph

in dendrogram form, derived from the Greek word δευδρος for tree. Further details can be found in Paelinck and Nijkamp (1975, pp. 203–207).

16.1.5 Interwovenness

Interwovenness may be defined as the spatial mixing of units belonging to different activities; one possible method to measure it is the following:

Divide an area into subareas (e.g., lattices) in number equal to *minimum* n_k, where the n_is are the number of plants in each sector k; this opens the possibility of having all K sectors represented. Note that this analysis is local, but the area as a whole could also be considered.

As an intermediate step, a Gini or other concentration coefficient could be computed for the sectors present in each subarea and compared to that of the expected concentration; the expected number of l-plants is n_l/n_k, n_k being the minimum number referred to in the preceding discussion. Moreover, subarea specialization coefficients might be considered, computed as $n_{kr}/E(n_{kr})$, where $E(n_{kr})$ is the expected value of k-plants in subarea r, as previously defined.

In order to measure interwovenness, one needs to set up a distance matrix between plants, then *count*, in terms of increasing distances, the first similar plant encountered, then the second, and so on; next, these counts need to be averaged over the (sub)-area. Call this index I_{ik} (i = degree, k = sector) and its average over all plants I^*_{ik} . If the latter indices exceed $K - 1$, $2(K - 1)$,. . ., interwovenness is largely present; otherwise, plants belonging to the same sector tend to cluster. If some k-plants tend to occupy systematically border positions, the first index (I_{k1}^*) would tend to exceed $n_r - K + 1$, where N_r is the total number of plants present in subarea r.

As an example, take $K = 3$, $N = 9$, and three plants per sector; lower critical values are 2 and 4, and the upper one is 7. With a random distance matrix, $I_1 = [5; 2.67; 4.67]$ and $I_2 = [7; 5; 7.67]$; plants of type 2 tend to concentrate a little more than the other types. For a distance matrix with a concentration pattern, the outcome was: $I_1 = [1, 1, 1]$ and $I_2 = [2, 2, 2]$.

16.2 Spatial Econometrics

Meta-analyses of different methodological approaches can be revealing.

The present chapter compares results from two methods to regroup spatial units—connectropy and cluster analysis—the resulting groupings then being compared to traditional macro-regions. An attempt also is made to discover reasons for matching or diverging results. Belgian and Dutch regional units are explored.

The next subsection summarizes developments of the respective methodologies, with Sect. 16.2.2 presenting the main results. Conclusions follow.

16.2.1 Methodology

This section summarizes two clustering methodologies: connectropy and cluster analysis.

16.2.1.1 Connectropy (Kaashoek et al. 2004)

Consider an axiomatic definition of nearness or proximity, along the lines of that of a distance measure or metric. Take any set of elements, S, and consider three of those elements, x, y, and z. Define on two of them a numerical function, $n(.,.)$, which obeys the following properties:

$$n(x,y) = 1, \quad x = y, \tag{16.3}$$
$$0 < n(x,y) < 1, \quad x \neq y, \tag{16.4}$$
$$n(x,y) = n(y,x), \tag{16.5}$$
$$n(x,y) \bullet n(y,z) \leq n(x,z). \tag{16.6}$$

Let function $n(.,.)$ be called a nearness measure. The similarity with an axiomatic definition of a metric should be obvious; now the well-known triangular inequality is stated in terms of a product and a $\leq$-relation.

A possible and useful member of the family of nearness (or closeness) measures is the following:

$$n(i,j) = \exp(-d_{i,j}), \tag{16.7}$$

where d_{ij} is any proper distance measure between entities i and j (e.g., the degree of contiguity), which is indeed a proper distance measure and which is used subsequently in this chapter. One can easily check that Eq. (16.7) is a proper closeness measure, given the exponential additivity property and the triangular inequality property of metrics.

Now consider the function

$$c_{ij} = d_{ij} \exp(-d_{ij}) - n(i,j) \bullet \ln n(i,j), \tag{16.8}$$

of which the similarity with an entropy measure is again obvious; indeed, read probability for variable n, and the formal likeness is complete. But this also is true from an intrinsic point of view; Eq. (16.8) indeed can be given an informational content. The occurrence of a long distance relation is relatively rare, so its observation contains a high degree of information, which is measured here by d_{ij}; in contrast, distance interaction confirms that the effective intensity of interaction declines with distance [a la Eq. (16.7), in a negative exponential way, similar to radioactive decay], and as entropy measures expected information, Eq. (16.7), which we call connectropy, measures the expected intensity.

Results of such an entropy-type measure can be treated as follows: suppose one avails of a matrix of connectropy measures, one then could try to extract groups with certain values of total connectropy, $\sum_{i,j} c(i,j)$ (to be compared with total entropy), such as declining levels from a maximum to a minimum; this specification amounts to applying linear assignment to the matrix, as will be shown next (based on Paelinck 2000).

Let **A**, of order $n \times n$, be that matrix; note that if the diagonal is not void, its terms have been put equal to zero in all the exercises performed. One could try and generate several paths, possibly generating cyclical groups (or circuits), in the sense that the last entity connects with the first one, each maximizing the sum total of the interrelationships present, the objective being to isolate *several* groups with declining maximal total internal intensity, and study the resulting interconnections. The *initial* mathematical program can then be written as

$$\max_{\mathbf{x}} \ \varphi \triangleq [\mathbf{vec}(\mathbf{A}')]'\mathbf{x} \tag{16.9}$$

subject to

$$\mathbf{Jx} \leq \mathbf{i}, \tag{16.10}$$

$$\hat{\mathbf{x}}\mathbf{x} = \mathbf{x}, \tag{16.11}$$

where **i** denotes the unit column vector; **vec(A′)** is the vectorization of the transpose of **A**, all diagonal elements omitted, so of order $n(n - 1)$ x 1; **x** is a vector of order n $(n - 1)$ x 1 of binary variables [conditions (16.11), where $\widehat{\mathbf{x}}$ is the diagonal matrix constructed from **x**]; and matrix condition (16.10) is a weakened assignment condition, matrix **J** being binary and of order $2n \times n(n - 1)$—it is the so-called assignment matrix, assigning a unique place to each of the units taken up in the solution.

Given this context, if the weak inequalities were to be replaced by equalities, exactly *n directed* relations would be selected, each agent appearing twice, so relaxation allows for generating *incomplete groups*. Meanwhile, the mathematical program (16.9) through (16.11) is the inverse (a maximum instead of a minimum) of the traveling salesman problem, but here the traveling salesman is in no hurry at all, and not obliged to pass through all of the potential nodes, and, moreover, s/he is allowed to go back and forth and to travel along disconnected paths!

Once the solution to the mathematical program has been obtained (by simple linear programming), one has generated a first group with maximal internal cohesion, meaning here total connectropy. One then cancels the corresponding entries of matrix **A**, rendering a matrix **A*** to be treated in the same manner, and so on until all of the matrix entries have been exhausted.

Symmetry of matrix **A** leads to symmetrical flows in the optimal solution, which can be prevented by introducing the conditions

$$x_{ij} + x_{ji} \leq 1, \quad \forall i < j, \tag{16.12}$$

which still do not guarantee complete connectivity. Such a requirement would lead to a quadratic assignment problem.

16.2.1.2 Clustering (Paelinck 2004)

The model here runs as follows: specify an objective function, to be minimized, as

$$\varphi = \Sigma_{i,j>i} d_{ij}\left(x_{i1}x_{j1} + \cdots + x_{ik}x_{jk}\right), \tag{16.13}$$

where i and j are indices for n observations and d_{in} entris of a distance matrix, $\boldsymbol{D}$, derived from the values of the observations. The x_{ic}, x_{jc}, $1 \leq c \leq k$ are binary decision variables assigning observations to k regimes. A value d_{ij} is only effective if both observations i and j belong to the same regime.

If **X** is the *kxn* matrix constructed from the row vectors of the x_{ic}'s, a first constraint is

$$\mathbf{Xi} = \mathbf{i}, \tag{16.14}$$

the ***I***'s being conformable unit column vectors. The meaning of this expression is that each x_{ic} should be allocated to one single regime. The next constraint is

$$\mathbf{X}'\mathbf{i} \geq \mathbf{I}, \tag{16.15}$$

meaning that each cluster should contain at least one observation. One extra constraint could be

$$x_{ic} = 1, \tag{16.16}$$

meaning that observation i belongs to cluster c, where i and c are arbitrary.

Each product in Eq. (16.13) can be linearized as follows: take, for example, the product $x_{ic}x_{jc}$ and write

$$d_{ij}\left(x_{ic} + x_{jc} \quad - 1\right) + y_{ijc} \geq 0, \tag{16.17}$$

where y_{ijc} is a real auxiliary variable. All possible values of the product $x_{ic}x_{jc}$ can be matched by inequality (16.17). Indeed, if both x_{ic} and x_{jc} equal one, the value of the left-hand side of inequality (16.17) is one, because y_{ijc}, appearing in the objective function to be minimized, is zero. If either x_{ic} or x_{jc} is zero, then the value of the left-hand side of inequality (16.17) is zero, y_{ijc} again being zero. Finally for both x_{ic} and x_{jc} equal to zero, the left-hand side of inequality (16.17) again is zero, now with $y_{ijc}= d_{ij}$, the smallest value it can take for expression (16.17) to be nonnegative.

The disadvantage of the specification just presented is that it introduces extra y_{ijc} variables (but those also are present in other specifications). However, this specification is useful because it can be solved by linear programming (specifying $x_{ic} \geq 0, \forall i, c$) and can be generalized to quadratic assignment problems.

In time series, continuity restrictions have to be imposed; in spatial analysis, contiguity restrictions are to be introduced in order to generate continuous regions. These latter restrictions are

$$\mathbf{x}_c'\mathbf{C}_1\mathbf{x}_c \geq \mathbf{i}'\mathbf{x}_c - 1, \tag{16.18}$$

where $\mathbf{C_1}$ is the contiguity matrix of order one. Indeed, the left-hand term is the sum of the relevant cross-products for cluster c, and that number should at least be equal to the total number of terms minus one.

Determining a linear analogue to inequality (16.18) has, as yet, not been possible, so a differential solution procedure was devised. For problems of the type investigated, a simple solution procedure can be constructed, the distances appearing in the distance matrix **D** being natural distances (i.e., distances along a line). Thus, classifying the items to be investigated from smallest to largest values, nondecreasing series are generated. Relative increases are present in terms of rows and columns of matrix **D**. Inspection of the column sums of the values above the main diagonal of matrix **D** show jumps at certain points; from there on, a new cluster should be investigated.

As to contiguity condition (16.18), it can be treated in the same way, starting from the smallest admissible (contiguous) values; then the procedure is same for the permutated matrix.

An index, I_c, can be computed as ϕ/ϕ_c , where ϕ is the value of the free optimum, ϕ_c being the contiguity-constrained one. This index is an indicator of the strength of the contiguity condition, a value of *1* meaning that the constraint is superfluous, and a value $I_c < 1$ measuring its strength. This index can be normalized so that it lies between 0 and 1, by computing an absolute maximum for ϕ_c, which would result from the free inverse (maximizing) problem; this adjustment has not been done in this study.

16.2.2 Applications and Comparisons

This section presents two applications—to The Netherlands and to Belgium—in terms of regional products and based on the methodology of Sect. 16.1.2. Traditionally, The Netherlands is divided into four macro-regions, namely North, East, South, and West; Belgium is considered in terms of three macro-regions, namely Flanders, Wallonia, and Brussels. Hence, the number of clusters to be considered in each case is four and three; the proof of the pudding lies in the results obtained to be compared to the prevailing traditional regional aggregation.

Comparisons with results from a connectropy analysis are presented afterwards.

16.2.2.1 The Netherlands

The relevant quantities concerning GRPs and contiguity degrees appear in Kaashoek et al. (2004). Table 16.3 lists the Dutch provinces portrayed in Fig. 16.3.

Using the provincial identification numbers appearing in Table 16.3, the nondecreasing order of the provinces is given by [5, 3, 10, 2, 1, 4, 12, 7, 6, 11, 8, 9]. The resulting, already reduced column rows are [0, 77, 79, 232, 508, (1115); 0, 47, 169, 845, (2056); 0, 143; 0] (original quantities are multiplied by 10), generating the four target clusters. The economically most active province, Zuid-Holland, is isolated; $\varphi = 2100$.

The contiguity constrained order of these provincial identification numbers is [3, 2, 5, 1, 4, 7, 6, 12, 11, 10, 8, 9] with (reduced) column sums [0, 52, 206, 442, 870, (1490); 0, 225, 347, (1420); 0, 881 (1167); 0, 551} resulting this time in isolating Zeeland, the third province in the nondecreasing order; $\varphi_c = 3570$, and $I_c = 0.5547$.

Table 16.3 The Dutch provinces

1. Groningen	2. Friesland	3. Drenthe
4. Overijssel	5. Flevoland	6. Gelderland
7. Utrecht	8. Noord-Holland	9. Zuid-Holland
10. Zeeland	11. Noord-Brabant	12. Limburg

Fig. 16.3 The Dutch provinces

16.2.2.2 Belgium

The quantities again are taken from Kaashoek et al. (2004), with Table 16.4 and Fig. 16.4 paralleling the preceding Tables 16.3 and Fig. 16.3 for the Netherlands.

The Belgian regions are analogous to provinces, except for those with provincial identification numbers in Table 16.4 of 1 and 12, the last one grouping the European Union, foreign embassies, and NATO, these entities being located inside the region Bruxelles-Capitale. A fictitious maximal degree of contiguity four has been assigned to this latter group.

Table 16.4 The Belgian provinces

1. Region of Bruxelles-Capitale	2. Antwerpen	3. Limburg
4. Oost-Vlaanderen	5. Vlaams Brabant	6. West-Vlaanderen
7. Brabant wallon	8. Hainaut	9. Liège
10. Luxembourg	11. Namur	12. Extra-territorial units

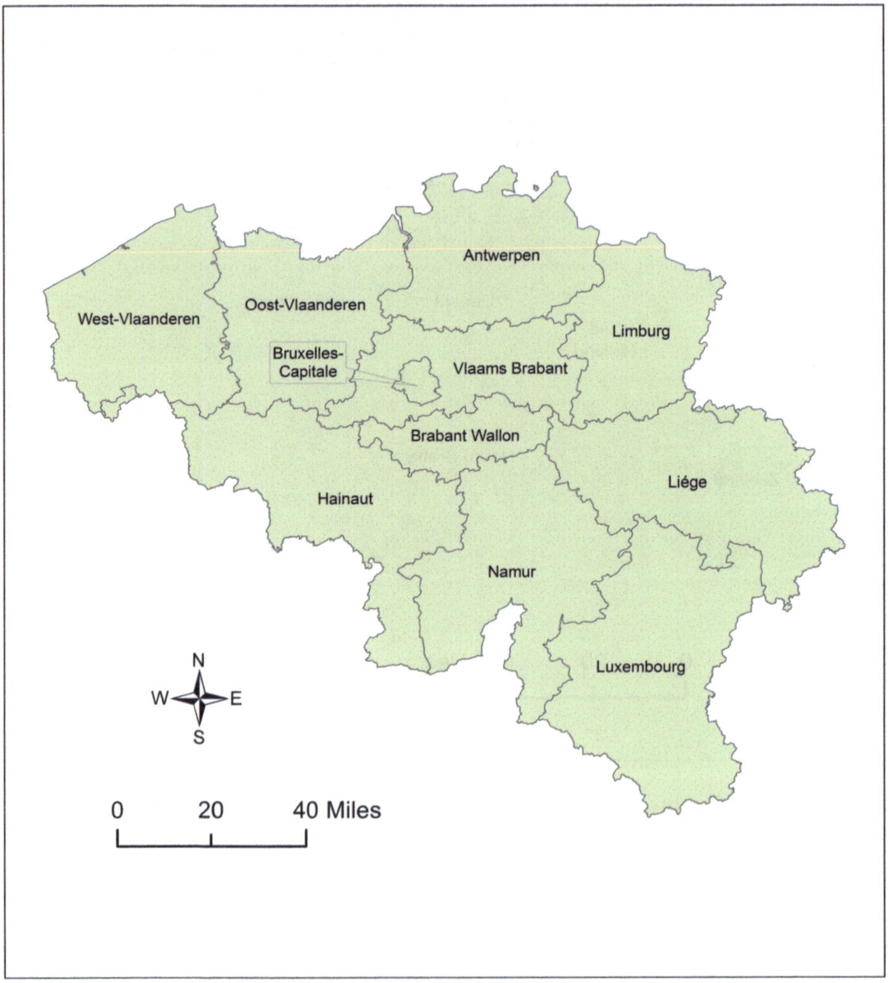

Fig. 16.4 The Belgian provinces

The nondecreasing order of the provincial identification numbers appearing in Table 16.4 is [12, 10, 7, 11, 3, 5, 8, 9, 6, 4, 1, 2], and the reduced column sum vector is [0, 19, 31, 52, (212); 0, 11, 13, (125); 0, 10, 50, (263); 0] (original quantities multiplied by 1000), yielding $\varphi = 199$, and isolating the province of Antwerpen.

As to contiguity, a special case was present as mentioned previously, the extra-territorial units, which had to be isolated from the very start of the analysis. The resulting classification is [12, 5, 8, 9, 3, 1, 7, 11, 10, 6, 4, 2], and the reduced column sum vector is [0; 0, 1, 29, 63, 245, (386); 0, 7, 19, (167); 0, 10, 192], with φ_c=*0.5660* and i_c= *0.3566*, a lower value than that obtained for the Netherlands.

16.2.2.3 Comments

The general outcome is that a much larger dispersion in the final contiguity-constrained clusters exists than is the case in official groupings. In the Netherlands, the first cluster groups five provinces as the North (Drenthe, Flevoland, Friesland, Groningen, Overijssel) and three as the Center-South (Gelderland, Limburg, Utrecht), with the West including only North- and South-Holland (omitting Utrecht, which traditionally belongs to the so-called Rimcity), and Zeeland joining Noord-Brabant.

The same type of characterization applies to Belgium, with a large central-south-east cluster (five units: Bruxelles-Capitale, Hainaut, Limburg, Liège, Vlaams-Brabant), a three-unit south-east cluster (Brabant Wallon, Luxembourg, Namur), and a three-unit north-west cluster (Antwerpen, Oost-Vlaanderen, and West-Vlaanderen, which thus excludes Vlaams-Brabant and Limburg).

Of course, using an indicator other than the one employed here (i.e., GRP), other results could be expected. Because comparing the results obtained here with the ones obtained by Kaashoek et al. (2004) using connectropy analysis would be interesting, the main results of another paper (Paelinck 2004) are presented next.

16.3 Comparison of Results

Table 16.5 reproduces groupings obtained by connectropy analysis; between parentheses one finds these results diverging from the previous ones.

Table 16.5 A comparison of cluster analysis and connectropy results

(a) The Netherlands
1. Overijssel, Utrecht, Gelderland, Limburg;
2. Noord-Holland, Noord-Brabant, (+Zuid-Holland);
3. Groningen, Zeeland, Flevoland, Drenthe, Friesland.
(b) Belgium
1. Bruxelles-Capitale, Oost-Vlaanderen, West-Vlaanderen, (+Antwerpen);
2. Limburg, Liège, Hainaut, Vlaams-Brabant;
3. Brabant Wallon, Namur, Luxembourg, (-Extra-territorial Units).

Of note is that the earlier 2004 results are not contiguity-corrected, as connectropy analysis aims at selecting families of regions based on the intensity of the economic links uniting them. Nevertheless, that the results diverge very little is remarkable and reveals the existence of similar clusters of elementary regions based on levels and interactions.

16.4 Conclusions

Apparently combining techniques from spatial econometrics and spatial structuring allows one to produce interesting mongrel methods. This consequence once more illustrates the virtues of regional science as a fundamentally interdisciplinary approach (Paelinck 1982).

Much remains to be done in the field just presented. More experience with observed data should reveal whether the potentials described in this chapter lead to efficient analytical tools.

References

Kaashoek, J., Paelinck, J., & Zoller, H. (2004). On connectropy. In A. Getis, J. Mur, & H. Zoller (Eds.), *Spatial econometrics and spatial statistics* (pp. 217–231). London: Palgrave.

Paelinck, J. (1982). Operational spatial analysis: Presidential address. *Papers and Proceedings of the Regional Science Association, 50*, 1–7.

Paelinck, J. (2000). *Controlling complexity in spatial modelling* (Working Paper No. 362). Munich: CESifo Group.

Paelinck, J. (2004). Transport and spatial clustering. In D. Hensher (Ed.), *Handbook of transport geography and spatial systems* (Vol. V, pp. 97–109). Amsterdam: Elsevier.

Paelinck, J., & Nijkamp, P. (1975). *Operational theory and method in regional economics*. Farnborough: Saxon House.

Paelinck, J., & Vossen, P. (1983). Axiomatics of conflict analysis. In W. Isard & Y. Nagao (Eds.), *International and regional conflict* (pp. 33–52). New York: Ballinger.

Chapter 17
Linear Expenditure Systems and Related Estimation Problems

Although linear expenditure systems (LES) belong to the realm of demand theory, they can be shown to be useful in other contexts; this demonstration is done in the next section, which presents possible alternative specifications in a spatial econometric surrounding. Sections 17.2 and 17.3 are devoted to estimation problems as they are encountered.

The morphism used in this chapter is a formal specification of demand theory.

17.1 Linear Expenditure Systems (Paelinck 1964; Solari 1971)

This section summarizes estimation results for three models, namely the level, growth rate model 1 (GRM1), and growth rate model 2 (GRM2) specifications.

17.1.1 Level Specification

Consider the following specification:

$$p_{it}q_{it} = p_{it}c_i + a_i\left(\sum\nolimits_i p_{it}q_{it} - \sum\nolimits_j p_{jt}c_j\right), \tag{17.1}$$

with $\sum_i a_i = 1$, where p_{it} denotes prices, q_{it} denotes real GRPs, and c_i and a_i are parameters. This model has been fitted to Belgian data (Source; EUROSTAT; years 1997–2008), with the results reported in Table 17.1; Fig. 17.1 portrays the relevant spatial units.

D. A. Griffith, J. H. P. Paelinck, *Morphisms for Quantitative Spatial Analysis*, Advanced Studies in Theoretical and Applied Econometrics 51,
https://doi.org/10.1007/978-3-319-72553-6_17

Table 17.1 Estimation results of the level model

Spatial units	c	a	c/μ (q)
1: Bruxelles Capitale	42,049	0.1827	0.8454
2: Antwerpen	41,238	0.1889	0.8386
3: Limburg	14,255	0.0632	0.8449
4: Oost-Vlaanderen	25,474	0.1241	0.8306
5: Vlaams-Brabant	22,208	0.1144	0.8227
6: West-Vlaanderen	22,666	0.0989	0.8457
7: Brabant Wallon	6633	0.0458	0.7786
8: Hainaut	18,335	0.0715	0.8605
9: Liège	16,813	0.0624	0.8668
10: Luxembourg	3767	0.0162	0.8556
11: Namur	6559	0.0319	0.8346

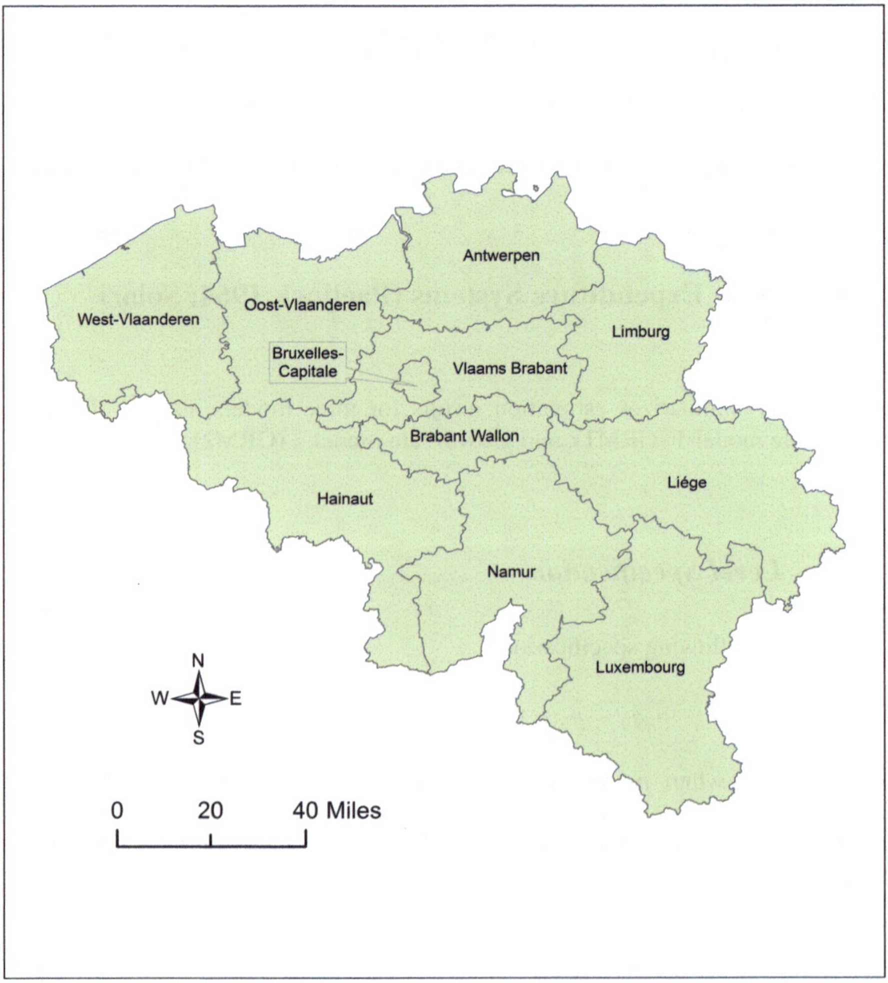

Fig. 17.1 Belgian spatial units

The estimation method, SDLS, is, as mentioned previously, further developed in Sect. 17.2.

Remarkably, c_i is nearly always equal to 0.80–0.85 of the average of the real GRPs, except for Brabant Wallon; no correlation was observed between that average and the a_is. The multiple determination coefficient is $R^2 = 0.9996$; all c_is and a_is are positive and highly significant.

17.1.2 Growth Rate Model 1

This model is specified as

$$r_{it} = c_i + a_i\left(r_t - \sum_j w_{jt}c_j\right), \tag{17.2}$$

where r^{it} denotes the observed growth rate between years $t - 1$ and t for region i, r_t denotes the overall growth rate, w_{jt} denotes the average GRP shares of periods $t - 1$ and t, and c_i and a_i again being the relevant parameters. Table 17.2 reports the estimation results for this model.

This time the c_is spatially are practically equal to the average growth rate; $\sum_i a_i$, though not constrained, is equal to 0.9821, and $R^2 = 0.8198$, largely significant for $F(99, 21)$, which is remarkable as first differences have to be explained. Again, all c_is and a_is are positive and highly significant.

Table 17.2 Estimation results for the GRM1

Spatial unit	c	a	$c/\mu(r)$
1: Bruxelles Capitale	0.0347	0.0996	0.9995
2: Antwerpen	0.0434	0.1064	0.9995
3: Limburg	0.0371	0.0977	0.9995
4: Oost-Vlaanderen	0.0371	0.0924	0.9997
5: Vlaams–Brabant	0.0414	0.0728	0.9997
6: West-Vlaanderen	0.0353	0.0941	0.9996
7: Brabant Wallon	0.0532	0.1008	0.9997
8: Hainaut	0.0325	0.0683	0.9996
9: Liège	0.0305	0.0750	0.9996
10: Luxembourg	0.0328	0.0919	0.9995
11: Namur	0.0385	00832	0.9996

Table 17.3 Estimation results for the LVLES

Spatial units	c	a	$c/\mu(r)$
1: Bruxelles Capitale	0.002118	0.0624	0.0610
2: Antwerpen	0.001725	0.1330	0.0502
3: Limburg	0.004546	0.0819	0.1224
4: Oost-Vlaanderen	0.004494	0.1457	0.1211
5: Vlaams-Brabant	0.008787	0.1037	0.2123
6: West-Vlaanderen	0.002676	0.0463	0.0758
7: Brabant Wallon	0.020597	−0.0945	0.3871
8: Hainaut	−0.000123	0.0875	−0.0038
9: Liège	−0.002117	0.0152	−0.0694
10: Luxembourg	−0.002119	0.0793	0.0646
11: Namur	0.005915	0.0932	0.1535

17.1.3 Growth Rate Model 2

This specification is

$$r_{it} = c_i + a_i \sum_i q_{it}(r_t - c_i), \tag{17.3}$$

where the q_{it} quantities have been divided by 10^3, and which might be called a Lotka–Volterra linear expenditure system (LVLES), because growth rates and levels of variables are jointly analyzed. Table 17.3 reports the estimation results for this model.

The remarkable feature here is that some of the c_is and the a_is are negative [for an interpretation of this outcome in demand theory, see Paelinck (1964, pp. 36–37) and Solari (1971, pp. 55 a.f.)]. The unconstrained $\sum_i a_i = 0.7614$, $\sum_i c_i = 0.0804$, and $R^2 = 0.7872$ again are highly significant (as are all coefficients c_i and a_i), which is of note again because first differences have to be explained.

17.1.4 Conclusion

That the model described by Eq. (17.1), the level one, is practically an interpolation, is not surprising because LESs have shown the same feature (Paelinck 1964). More striking are the results of GRMs 1 and 2, especially the latter, because it is a true dynamic model, fit for projections and dynamic policies.

17.2 Different Estimators Compared

To begin with, the special estimator SDLS, used in Sect. 17.1, displays some apparent common features with other estimators that link them together. This section addresses this aspect and submits it to analysis. A common family feature is that all of these estimators are least distance ones.

17.2.1 Simultaneous Dynamic Least Squares

The origin of the idea for this estimator may be found in the construction of a dynamic nonlinear macro-model for Belgium (de Biolley and Paelinck 1970), for which task an appropriate projection estimator was investigated; the result was SDLS estimation.

The motivating idea was the following: develop an estimator that minimizes some distance between, not the computed and observed values, but the *dynamically simulated* and the observed ones. Moreover, if a dynamic process is Markovian, then an optimal initial value (i.e., starting point) for a simulation also could be computed as an extra parameter. Table 17.4 shows the results of running a program for a one-variable Markov process ($y_t = ay_{t-1} + b$).

Initially, following Paelinck (1990), the simulated values were obtained by iteration, but the program, as it stands now, generates them together with the parameters.

The program can be generalized to Markov systems of equations, as shown in Griffith and Paelinck (2011, Chap. 11). Another possibility is to take account of further lags, as developed hereafter. The Markov process may be generalized to a Yule process as follows:

$$y_t = xay_{t-1} + (1 - x)ay_{t-2} + b, \tag{17.4a}$$

$$0 \leq x \leq 1, \tag{17.4b}$$

so only a fraction x of y_{t-1} affects y_t, the remaining fractional effect being allotted to y_{t-2}. Table 17.5 shows the result for an arbitrary time series that can be read from

Table 17.4 Results for a one-dimensional Markov process

Parameters and optimal initial value	Endogenously generated variables	Simulation equations	Observed and simulated value difference	Sum of squares
0.965759179	4.820458406	1.69484E-07	0.179541594	1.937840856
1.536278211	6.191680437	2.74248E-07	−0.19168044	
3.400620049	7.515950737	3.12043E-07	−0.51595074	
	8.794876918	2.94736E-07	0.205123082	
	10.03001155	2.31687E-07	0.969988447	
	11.22285406	1.29685E-07	−0.22285406	
	12.37485253	−5.6892E-09	−0.37485253	
	13.48740545	−1.7095E-07	−0.48740545	
	14.56186346	−3.6403E-07	0.438136537	

Table 17.5 Results for a generalized Markov process

Parameters and optimal initial value	Endogenously generated variables	Simulation equations	Observed and simulated value difference	Sum of squares
1.102912	9.920109	5.42E-07	−0.73558	6.277511
2.323418	9.365172	3.27E-07	−0.34763	
0	13.26442	4.34E-07	0.952902	
	12.65237	1.65E-07	0.277869	
	16.9529	0	1.27632	
	16.27787	1.78E-07	−0.72353	
	19.27632	−2.3E-07	−1.41651	
	20.27647	4.04E-07	0.68657	
	23.58349			

the table, of which the interpretation is identical to that for Table 17.4, except that the third term is now x, and that both y_{t-1} and y_{t-2} are optimized as starting values.

The result is extreme, in the sense that the optimal adjustment implies a complete second-order lag, with $R^2 = 0.974$ (the first-order lag produces a value of 0.956).

17.2.2 Reduced Form and Two Stage Least Squares Estimation

When the system of equations is static, SDLS estimators are identical to reduced form ones.

In two equations and two endogenous variables, y_1 and y_2, the simulation system can be written as follows, where Pareto circlets denote endogenous SDLS variables:

$$y_1^o - ax^* - b - cy_2^o = 0, \tag{17.5a}$$

$$y_2^o - \alpha z^* - \beta - \gamma y_1^o = 0, \tag{17.5b}$$

the estimates being obtained by minimizing a function on d_l and d_2 of Eqs. (17.7a) and (17.7b). Hereafter, x^* and z^* are exogenous, with the system being identified as

$$d_1^2 + d_2^2. \tag{17.6}$$

$$y_1 - y_1^o = d_1, \tag{17.7a}$$

$$y_2 - y_2^o = d_2. \tag{17.7b}$$

From Eqs. (17.5a) and (17.5b), one derives

$$y_1^o = D^{-1}[ax^* + c\alpha z^* + (b + c\beta)], \tag{17.8a}$$
$$y_2^o = D^{-1}[\gamma ax^* + \alpha z^* + (\beta + \gamma b)], \tag{17.8b}$$

with D^{-1} the inverse denoting determinant of the system. But these are the RF equations from which the structural parameters can be derived. For two-stage least squares (2SLS), the equations are as follows:

$$y_1 = ax^* + b + cy_2^o, \tag{17.9a}$$
$$y_2 = \alpha z^* + \beta + \gamma y_1^o, \tag{17.9b}$$

with

$$y_1^o = dx^* + ez^* + f, \tag{17.10a}$$
$$y_2^o = \delta x^* + \varepsilon z^* + \varphi. \tag{17.10b}$$

However, this substitution does not yield the same values for the sextet $\{a, b, c; \alpha, \beta, \gamma\}$. This equivalency is the case only if these estimates are computed from Eqs. (17.8a) and (17.8b) as in RF estimation.

17.2.3 Latent Variables

One can ask the question whether or not a series of latent variables can be generated endogenously like in the SDLS estimation process.

Consider the following model:

$$y_i = x_i + az_I^* + b + e_i, \tag{17.11}$$

where x_i denotes an unknown series of latent variables (see Bartholomew and Knott 1999), z_I^* denotes an exogenous variable, and e_i denotes the computed residuals. The coefficient of x_i is set equal to one, because any other coefficient would affect the series itself by an inverse multiplicative factor.

Computing the parameters of Eq. (17.11) without restrictions yields

$$x_i = y_i \forall i, \tag{17.12a}$$

$$a, b, e_i \forall i = 0. \tag{17.12b}$$

To prevent this outcome, the following constraints can be imposed:

$$\sum\nolimits_I e_i^2 = t. \tag{17.13a}$$

where t should lie between zero and its value from an estimation without the x_i-series, and

Table 17.6 Estimation results with an endogenous series of latent variables

Parameters	Vector of latent variables	e_is	Condition (17.12b) and the value of t
1.03675128	0.602826296	−0.10646783	1.17747E-07
0.503641534	0.120882243	−1.73477762	13.18999987
	0.093496931	−2.74414358	
	0.561854092	0.861001814	
	0.54439564	0.824250534	
	1.540693053	−0.22809099	
	1.005105439	−0.72925465	
	2.698026492	0.651326854	
	1.672673888	−0.43357438	
	2.351085399	−0.14873717	

$$\sum\nolimits_{I} e_i x_i = 0, \tag{17.13b}$$

to prevent x_i from reproducing the e_is. Table 17.6 presents the resulting estimations.

The sum of squared residuals is fixed at half of the value of the initial regression without latent variables.

17.2.4 Linear Expenditure Systems

In former times, LES had to be computed by iteration (again Paelinck 1990); hence, the idea of borrowing the SDLS approach, generating endogenous variables.

A linear expenditure system—now taken from demand theory—can be specified as follows (see Sect. 17.1):

$$p_{it}^{*} q_{it} = p_{it}^{*} a_i + b_i \left(d_t - \sum\nolimits_{i} p_{it}^{*} a_i \right), \tag{17.14}$$

where the p_{it}^{*} are the (exogenous) prices of goods i at times t, $p_{it}^{*} a_i$ committed expenditures, and the b_i fractions of so-called supernumerary income spent on the different goods i. The following variables and parameters are taken up in a 2-good model solved by SDLS: q_{it}, s_t (supernumerary income), d_t (total expenditure), a_i and b_i, the latter summing to one, for consistency reasons of the LES; Table 17.7 shows the results.

The curious result of this estimation is that b_I is zero, so that only committed expenditures are present, an outcome due to the fictitious data used.

Table 17.7 Estimation results of a LES

					Eq. (17.14)					$b_1 + b_2 = 1$ & quadratic sum
a_1, a_2, b_1, b_2	q_{1t}	q_{2t}	d_t	s_t	$p^*_{1t}\ q_t$	$p^*_{2t}q_t$	s_t	d_t	$d_t - d_t(ob)$	of differences
3.14255	3.14255	3.42872	9.99999	4.64693	−1.2E-7	2.9E-8	2.0E-8	−2.8E-8	−4.9E-6	2.93E-8
1.10526	3.14255	4.42872	11.9999	6.64692	−1.6E-7	5.3E-8	9.0E-8	0	−5.8E-6	7.35E-10
8.06E-8	3.14255	3.85745	13.9999	5.50437	−1.7E-7	−1.8E-8	2.5E-10	2.6E-8	−7.0E-6	
1	3.14255	2.17872	14.9999	4.29385	−1.4E-7	−2.2E-8	3.2E-10	3.3E-8	−7.3E-6	
	3.14255	2.67872	16.9999	6.29385	−1.8E-7	1.3E-8	5.0E-10	−2.2E-8	−8.2E-6	
	3.14255	2.14309	17.9999	4.15130	−1.4E-7	−3.3E-8	7.6E-10	5.1E-8	−9.0E-6	
	3.14255	2.10745	20.9999	4.00875	−1.2E-7	−2.4E-8	−3.8E-10	−3.2E-9	−1.1E-5	
	3.14255	1.88596	21.9999	3.90349	−1.1E-7	−1.3E-8	0	9.5E-9	−1.1E-5	
	3.14255	1.38121	23.9999	1.65569	−8.2E-8	8.1E-9	−2.9E-8	−3.7E-9	−6.2E-6	
	3.14255	1.71454	25.9999	3.65568	−1.2E-7	5.2E-8	−4.0E-8	−8.2E-8	−1.3E-5	

17.2.5 Conclusion

The preceding situations just described show that there are quite a few models that can be treated along the lines of SDLS with simultaneously generated endogenous variables. One can expect other instances to turn up, in which a simple method, as described in the previous sections, can be used.

17.3 Distribution-Free Maximum Likelihood Estimation

Because all of the methods treated in the previous sections are distribution-free (i.e., they assume no specific frequency distribution), investigating the case of distribution-free maximum likelihood estimators seems in order.

Pseudo-maximum likelihood methods are presented elsewhere (see Greene 2008, pp. 511–517); other possibilities of distribution-free maximum likelihood are investigated here.

17.3.1 The Single Equation Case

The following three cases may be considered:

$$p_i\ \varphi|\varepsilon_i|, \tag{17.15}$$

the ε_i being the theoretical disturbances:

$$p_i\ \varphi\ \exp(-|\varepsilon_i|) \tag{17.16}$$

and

$$p_i\ \varphi\ \exp\left(-{\varepsilon_i}^2\right). \tag{17.17}$$

Maximizing the likelihood for Eq. (17.15) leads to minimizing $\sum_i \ln|\varepsilon_i|$. For Eq. (17.16), it leads to minimizing $\sum_i|\varepsilon_i|$ (see Taylor 1974, for developments), whereas for Eq. (17.17), it leads to OLS (see Tintner 1952, p. 27).

Table 17.8 summarizes results for the three specifications applied to the series generated by the following equation:

$$y_i = ax_i + b + \varepsilon_i. \tag{17.18}$$

Meanwhile, Table 17.9 presents analysis results for the data appearing in Table 17.8.

In this simple case, the results are practically equivalent, OLS, as expected, providing the largest R^2.

Table 17.8 Data generated with Eq. (17.18)

y_i	10	11	14	13	17	17	20	21	25	26
x_i	1	2	3	4	5	6	7	8	9	10

Table 17.9 Computational results

Parameter\equation	(17.15)	(17.16)	(17.17)
a	1.7362	1.8665	1.8061
b	8.2638	7.3350	7.4669
(Pseudo-)R^2	0.9590	0.9641	0.9666

17.3.2 Interdependent Systems

Here specifications (17.15) through (17.17) are applied to the following model:

$$y_i = ax_i + bu_i^* + c + \varepsilon_{1i}, \tag{17.19a}$$

$$x_i = dy_i + ez_i^* + f + \varepsilon_{2i}, \tag{17.19b}$$

the starred variables being exogenous. Table 17.10 presents the generated data series used.

The difference between this situation and the one-equational model is that the term $-\ln|\det A| = -\ln|1-ad|$ enters the log-likelihood expression in the minimizing procedure (Anderson 1958, pp. 10–11). Table 17.11 presents the results for Eqs. (17.15) through (17.17), first without (a), then with (b) the additional determinant term, except for Eq. (17.15) which has a very bad fit for Eq. (17.19b).

In terms of overall fit, both Eqs. (17.16) and (17.17) give admissible results, though (17.16) dominates (17.17) in terms of the R^2s.

To test an observed case, the data from Sect. 17.2 was used for the example presented here, neglecting the results obtained there, and now using the equation

$$\mathbf{Y} = \rho\mathbf{W}\mathbf{y} + \mathbf{X}\boldsymbol{\beta} + \boldsymbol{\varepsilon}, \tag{17.20}$$

Table 17.10 Generated data

y_i	x_i	u_i	z_i
10	10	1	5
12	10	1	5
14	11	1	5
15	12	2	4
18	13	2	6
20	13	2	6
25	14	3	6
27	15	3	7
30	15	2	7
32	16	4	7

Table 17.11 Computational results for the data appearing in Table 17.10

Equation\parameter	a	b	c	d	e	f	R^2 (16.19a)	R^2 (16.19b)
(17.15)a							0.9463	<0
(17.15)b							0.7317	<0
(17.16)a	0.5671	5.7316	0	0.2678	0	7.2557	0.9274	0.9554
(17.16)b	0.6666	5.3338	0	0.2721	0.0867	7.2454	0.9294	0.9558
(17.17)a	0.5680	5.9292	0	0.2679	0	7.4616	0.2866	0.9848
(17.17)b	0.5559	5.9927	0	0.2546	0.0633	7.3633	0.2948	0.9840

Table 17.12 Parameters estimates for model (17.20)

Equation\parameter	a	B	c	d	ρ	R^2
(17.15)a	−82537.3	95468.6	136018.2	−11636.7	0.4124	0.1742
(17.15)b	Idem					
(17.16)a	26773.0	177580.5	17833.9	−8366.5	0.1649	0.4921
(17.16)b	31036.8	2087840.6	185420.2	−7404.8	0.1259	0.4345
(17.17)a	−67875.1	115410.3	108552.2	131530.8	0	0.7027
(17.17)b	Idem					

with the spatial weights matrix **W** defined by the inverse degrees of contiguity corresponding to Fig. 17.1. Table 17.12 reports the results along the lines of Table 17.11, ρ being the parameter in Eq. (17.20).

Again, the results of Eq. (17.15) are very poor. Equation (17.16) performs much better, confirming the previous results. Finally, Eq. (17.17) reproduces practically the same results as at the end of Sect. 17.3.1. The method developed in that section seems to dominate the present approach.

17.4 Chapter Conclusions

Fifty years later (see Paelinck 1964), LES made a surprising comeback; the main conclusion of this chapter is that "far away and long ago" could be a motto in spatial econometric practice. In fact, formal equivalences have no frontier and often—as is the case here—lead to related fields.

The difference with the past is that new computational methods are available, which allows one to apply former specifications to much larger problems without using simplifications imposed by the lack of adequate computational means.

Other examples will certainly creep up, especially as these few examples of fruitful generalization demonstrate.

References

Anderson, T. (1958). *Introduction to multivariate statistical analysis*. New York: Wiley.

Bartholomew, D., & Knott, M. (1999). *Latent variable models and factor analysis*. London: Arnold.

de Biolley, T., & Paelinck, J. (1970). Un modèle dynamique pour la Belgique. In J. Paelinck (Ed.), *Optimisation et Simulation de Macro-décisions* (pp. 221–275). Namur: Economics Faculty, Namur University, Research Centre.

Greene, W. (2008). *Econometric analysis* (6th ed.). Upper Saddle River, NJ: Pearson-Prentice.

Griffith, D., & Paelinck, J. (2011). *Non-standard spatial statistics and spatial econometrics*. Berlin: Springer.

Paelinck, J. (1964). *Fonctions de consommation pour la Belgique 1948-1959* (Publications No. 1). Namur: Faculté des Sciences Economiques, Centre de Recherches, Namur University.

Paelinck, J. (1990). Some new estimators in spatial econometrics. In D. Griffith (ed.), *Spatial Statistics: Past, present and future* (Monograph No. 12) (pp. 163–181). Ann Arbor, MI: Institute of Mathematical Geography.

Solari, L. (1971). *Théorie des choix et fonctions de consommation semi-agrégées: modèles statiques*. Geneva: Droz.

Taylor, G. (1974). Minimizing the sum of absolute errors. In P. Zarembka (Ed.), *Frontiers of econometrics* (pp. 169–190). New York: Academic.

Tintner, G. (1952). *Econometrics*. New York: Wiley.

Chapter 18
Structural Indicators Galore...

This chapter treats and compares a certain number of what we dubbed structural indicators; these indicators concern the structure of spatial linkages. The two main categories studied hereafter are spatial discount functions (SDFs) and concentration indexes; they both obey a series of rich formalisms.

18.1 Spatial Discount Functions

SDFs have been used for over 50 years; some are even derived from the popular radioactive decay function:

$$p = \beta e^{-\alpha t}, \tag{18.1}$$

in which t is replaced by d, some distance measure; the left-hand side variable is denoted by p because often SDFs are used to compute potentials, sums of spatially discounted values. But p also could stand for probability, because the SDFs could equally well represent probability density functions (Ord 1972). This chapter establishes some other possibilities, beginning with an extension of the preceding function presented in Eq. (18.1).

18.1.1 The Tanner Function

The Tanner function (Tanner 1961) is expressed as

$$p = \beta d e^{-\alpha d}, \tag{18.2}$$

D. A. Griffith, J. H. P. Paelinck, *Morphisms for Quantitative Spatial Analysis*, Advanced Studies in Theoretical and Applied Econometrics 51,
https://doi.org/10.1007/978-3-319-72553-6_18

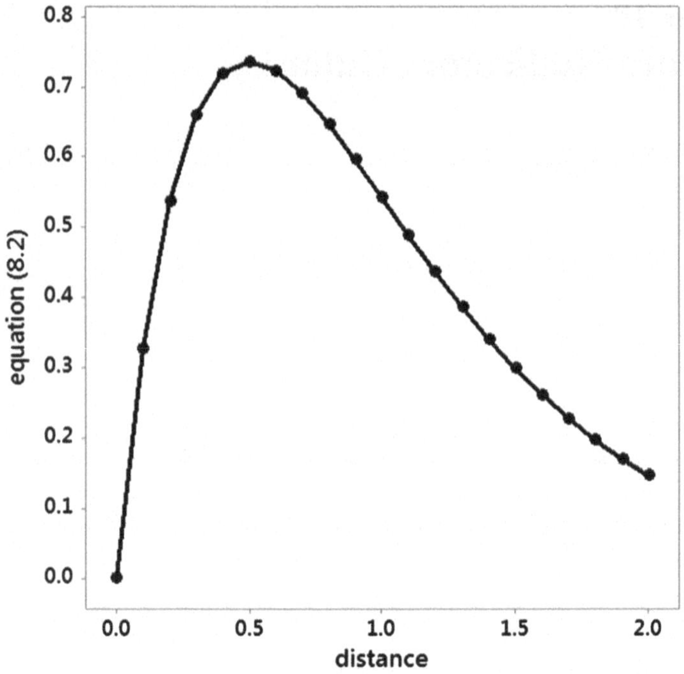

Fig. 18.1 A Tanner function

where p is the local discount at distance d. In its original version, this function was said to capture the fact that commuters-to-work prefer to have a job away from home at a reasonable distance, but the local density of commuters under the SDF given by Eq. (18.1) over a circular area is given by Eq. (18.2) times the distance differential, because the total potential has to be integrated over a two-dimensional area. If the behavioral assumption still holds, the second factor on the right-hand side of Eq. (18.2), d, should be squared. The integral-normalizing (total density) constant β in Eq. (7.2) can be computed as α^2; the same holds for the extended function just discussed.

Figure 18.1 portrays a typical case with $\alpha = 2$; for $\alpha \rightarrow \infty$, function (18.2) approaches function (18.1), and for $\alpha \rightarrow 0$, the function is increasing nearly everywhere; all empirical cases show a finite nonzero α (see Sect. 18.2).

Finally, noteworthy is that the maximum of Eq. (18.2) occurs at $d = \alpha^{-1}$, which equals 0.5 in Fig. 18.1; one can check from this that the previously mentioned properties of Eq. (18.2) hold. Furthermore, $p_{\max} = ae^{-1}$, which exceeds 1 for $\alpha > e$.

Table 18.1 Selected values of function (18.3)

Friction of distance parameter	Selected distances (in an arbitrary primary unit) between regions r and s					
α	0	0.25	0.67	1.50	4.00	∞	Maximum
0.00	0.000	0.000	0.000	0.000	0.000	1	∞
0.05	0.123	0.154	0.202	0.289	0.497	0	31.840
0.10	0.226	0.280	0.362	0.498	0.758	0	14.820
0.40	0.584	0.741	0.842	0.965	0.975	0	2.61
1.00	0.824	0.974	0.984	0.934	0.696	0	0.650
7.00	0.992	0.777	0.552	0.327	0.166	0	0.019
20.00	0.999	0.480	0.265	0.148	0.069	0	0.002
55.00	1.000	0.254	0.125	0.066	0.029	0	0.000
150.00	1.000	0.097	0.056	0.029	0.012	0	4.4×10^{-5}
400.00	1.000	0.044	0.025	0.012	0.005	0	6.2×10^{-6}
∞	1.000	0.000	0.000	0.000	0.000	0	0.0000

18.1.2 The Ancot–Paelinck Function

Ancot and Paelinck (1983a) start with Eq. (18.2) but replace d by a logarithmic function and the exponential by a linear one, obtaining the following final function:

$$p = \exp(1 - \alpha^*)[\ln(1 + \alpha d) + \alpha^*](\alpha d + 1)^{-1}, \tag{18.3}$$

where $\alpha^* = \alpha/(1 + \alpha)$, $\alpha \geq 0$. This function reaches its maximum of 1 (no integral normalization was computed; see the relevant subsequent discussion) at $d = (e^{1-\alpha^*} - 1)/\alpha$, so like for function (18.2), it involves a shifting coordinate. Table 18.1 (from Ancot and Paelinck 1983b, p. 238) reports selected values for it.

This function could be successfully used to compute, simultaneously, the spatial discount parameter, α (to derive potentials), a dynamic adaptation parameter, and all of the system parameters. As noted in preceding discussion, the spatial discount function is not integral-normalized (in fact, its integral over the interval $[0, \infty]$ is not convergent, being of $O\{\ln(d + 1)[\ln(d + 1) + 1]\}$, but a practical solution could be a division by the sum of computed weights, possibly in an iterative procedure.

18.1.3 The Continuous Poisson Function

Recall the Poisson density function

$$p = e^{-\alpha}\alpha^{-d}/d!, \tag{18.4}$$

where d is a nonnegative integer. For this function, $\sum(0,\infty) = 1$. In the Stirling approximation to $d!$ appears d^d, which suggests the following continuous version:

$$p = \beta(\alpha/d)^{d/\alpha} \tag{18.5}$$

Unlike the Tanner function, here $p \rightarrow \beta$ for $d \rightarrow 0$, its value for $d = 1$ is $\beta\alpha^{-1}$, and its value for $d = \alpha$ is β, its maximum occurring at $d = \alpha/e$ for $p = \beta e^e$; if $\beta = e^{-e}$, the maximum value of p is 1.

The integral-normalizing value of β cannot be computed, because function (18.5) does not seem to be analytically integrable. Nevertheless, this problem can be approached in the following way. Let us first render Eqs. (18.2) and (18.5) measurement neutral by substituting $\beta = \alpha$ into Eq. (18.2), and $\beta = n/\alpha$ in Eq. (18.5)—so the value of p for $n = 0$ is equal to zero. The idea now, taking logarithms of both functions, is to write for the net difference between ln[p(18.2)] and ln[p(18.5)]:

$$\text{Ln}[\alpha] - \alpha n \geq -\ln[\mu] + n\ln[\mu/\mu] - (1/\mu)n\ln[n], \tag{18.6}$$

because Eq. (18.2) is known to have a convergent density integral.

The right-hand side of inequality (18.6) has a maximum at $n = \mu/e$ equal to $-ln$ $[\mu]$, so condition (18.6) reduces to

$$\text{Ln}[\alpha] - \alpha\mu/e = -\ln[\mu], \tag{18.7}$$

which has a unique solution for any $\mu \geq 0$ [the right-hand member of Eq. (18.7) is finite, whereas the left-hand side is linear in α], establishing that Eq. (18.5) has a convergent density integral.

18.1.4 The Lognormal Function

A three-parameter, this time, lognormal function (Aitchison and Brown 1957) also could be a SDF; it presents itself as

$$p\sim N\left[\ln(d-\gamma),\mu,\sigma^2\right], \tag{18.8}$$

with γ being the starting value of the process, possibly equal to zero, thus reducing the parameters to be estimated to two, in which case the mode is 1; otherwise, $1+\gamma$.

18.1.5 The Loglogistic Function

All of the preceding SDFs are asymmetric; another possible function is the loglogistic, involving two parameters.

The frequency distribution function derived from the well-known logistic integral

$$l = \left(1 + \gamma e^{-\alpha \ln d}\right)^{-1} \tag{18.9}$$

is

$$p = \alpha\left(2 + \gamma^{-1} e^{\alpha d} + \gamma e^{-\alpha d}\right)^{-1} \tag{18.10}$$

for which the maximum occurs at $d = \alpha^{-1}\ln[\gamma]$, with the value of p equal to $\alpha/4$, which can be larger than 1.

18.1.6 Conclusions

Comparing the five SDFs based on a certain number of operational requirements is an interesting exercise. Ancot and Paelinck (1983a, b) list six such criteria, to which a seventh is added here. Table 18.2 reproduces all of them.

Starting with the Tanner function, various possibilities exist, and hence compromises can be made. If one rewrites function (18.2) as

$$p = \beta^* \alpha d e^{-\alpha d}, \tag{18.11}$$

then, if $\beta^* = e$, $p_{\max} = 1$. But the integral norm is not equal to one (it is variable, and equal to e/α), the choice of the distance unit being indifferent. If $\beta^* = \alpha$, then

Table 18.2 Operational requirements for SDFs

	SDF				
Requirement	Tanner	Ancot–Paelinck	Continuous Poisson	Lognormal	Loglogistic
1. $0 \leq p \leq 1$	Y–N	Y	Y	Y	Y
2. Max $p = 1$	Y–N	Y	Y	N	N
3. $p \rightarrow 0$, $d \rightarrow \infty$[a]	Y–N	Y	Y	Y	Y
4. $p = f$(1 parameter)	Y–N	Y	Y	N	N
5. Measurement neutral	Y–N	Y	Y	Y	N
6. Easy estimation	Y	Y	?	?	?
7. $\int_0^\infty$ convergent	Y	N	Y	Y	Y

[a]Except some special values of α

Table 18.3 Empirical values of α for the Ancot–Paelinck SDF

	Sector				
Variable	Food	Textile	Iron and steel	Machinery	Financial institutions
Demand t	3.71	1 09	0.97	1.08	4.02
Demand $t-1$	2.67	1.46	–	–	2.36
Supply t	1.42	–	–	10.92	–
Supply $t-1$	–	–	1.66	1.83	2.13

$p_{\max}= \alpha/e$, which is variable and generally unequal to one; the integral norm is equal to one, but the choice of the distance measurement unit is not indifferent.

The Ancot–Paelinck function has one drawback: its integral is not convergent. But $p_{\max} = 1$ is guaranteed, because it satisfies the remaining five requirements.

The continuous Poisson function has the first five requirements satisfied, but no empirical experience with it is known, because this function has been put forward only in this study; the convergence property of its integral could be established.

As for the lognormal function, its properties are well known; in particular, the variable d is mean-variance normalized. How troublesome the estimation of its two or three parameters could be for the computation of a space potential is not known.

Finally, the loglogistic function has a mixed score list; it could be improved as far as requirement 5 is concerned, by using a specification present in Eq. (18.3), to wit replacing $\alpha\ln[d]$ by $\ln[\alpha d]$. Nevertheless, requirements 2 and 4 would remain problematic.

To summarize, the Ancot–Paelinck specification has much to be said for it; Table 18.3 lists a few empirical values of α (from Ancot and Paelinck 1983a).

For this particular SDF, a large variety of values exist that can be given an empirical interpretation as a function of the characteristics of the sector and its locational pattern in the Europe of the Six (the Fleur model, for which the Ancot–Paelinck SDF was used, covered regions of Belgium, France, Germany, Italy, Luxembourg, and The Netherlands).

18.2 Dispersion Coefficients

The data used in Sect. 18.1 can be analyzed in an alternative way to disclose other aspects of multiregional growth. Tools to this end can be found throughout the classical nonspatial literature, as becomes clear from the ensuing discussion.

Table 18.4 presents growth rates (r = national; r_i denotes A, B, C, and D in Table 18.1) that are used throughout this section for illustrative purposes.

Table 18.4 Growth rates

Periods	r	r_1	r_2	r_3	r_4
2/1	0.035977	0.031949	0.030024	0.037634	0.039839
3/2	0.042169	0.038493	0.0318	0.041663	0.054895
4/3	0.046868	0.078745	0.050956	0.03384	0.052705
5/4	0.035852	−0.12236	0.057516	0.057454	0.070604
6/5	0.008577	−0.11603	0.025235	0.024784	0.025784
7/6	0.039484	−0.02644	0.037347	0.0495	0.050278
8/7	0.057208	0.063094	0.072226	0.045279	0.070662
9/8	0.062109	0.076619	0.069763	0.058731	0.056733
10/9	0.048989	0.085851	0.044311	0.044292	0.045447
11/10	0.045065	0.021376	0.056375	0.044342	0.049371
12/11	0.026758	0.012821	0.034323	0.030708	0.017985
13/12	0.033767	0.011902	0.030384	0.043014	0.025167
14/13	0.03266	0.011726	0.029488	0.04124	0.024549
15/14	0.034285	0.058841	0.030406	0.0302	0.035638
16/15	0.050857	0.041384	0.049403	0.043166	0.075266
17/16	0.057323	0.013359	0.058872	0.06622	0.055361
18/17	0.061063	0.020162	0.063635	0.061883	0.075227
19/18	0.078583	0.103206	0.078791	0.073285	0.080122
20/19	0.061265	0.084357	0.066361	0.059075	0.051802
21/20	0.04338	0.025324	0.037331	0.052311	0.035295
22/21	0.026443	0.038133	0.032473	0.024972	0.019575
23/22	0.024466	0.01146	0.019915	0.028605	0.024222
24/23	0.044323	0.064467	0.035892	0.044405	0.041885
25/24	0.046443	0.074436	0.060632	0.035681	0.047365
26/25	0.058134	0.034835	0.064332	0.060779	0.057528
27/26	0.038391	0.105145	0.037445	0.028496	0.030612
28/27	−0.03325	−0.08341	−0.02206	−0.02753	−0.03205

18.2.1 Variance Analysis

Figure 18.2 presents the data of Table 18.4 graphically.

At face value, nothing important seems to appear in Fig. 18.2; Fig. 18.3 seems to show that the variation coefficient, $v(\times 10)$, increases with the average growth rate, r (it fluctuates between −0.000860 and 0.002955, with an average of 0.000696).

A regression of the variation coefficient of the spatial growth rates on the overall growth rate reveals a homogeneous linear relation, the regression coefficient being $a = 0.0152(t = 2.8168, p = 0.0093)$, with $r^2 = 0.2409$ ($F = 7.9341$; $p = 0.0093$), implying that the spread of the spatial growth rates increases with the overall growth rate, which suggests that the fact that the dominating regions benefit more from overall growth than their minor partners.

Other links between overall and spatialized growth should be investigated, the methods just exposed showing the way.

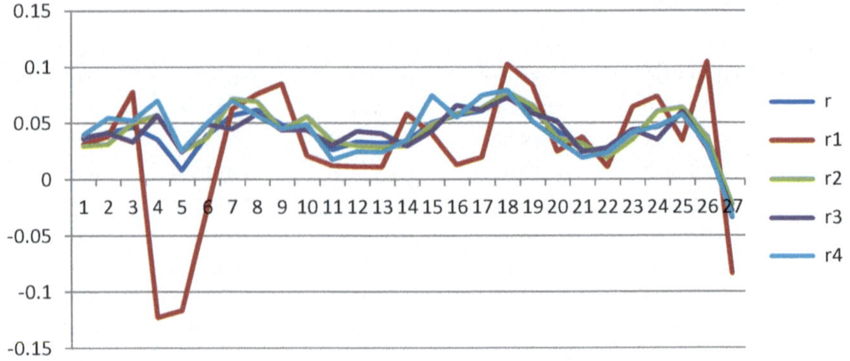

Fig. 18.2 Growth rates appearing in Table 18.4

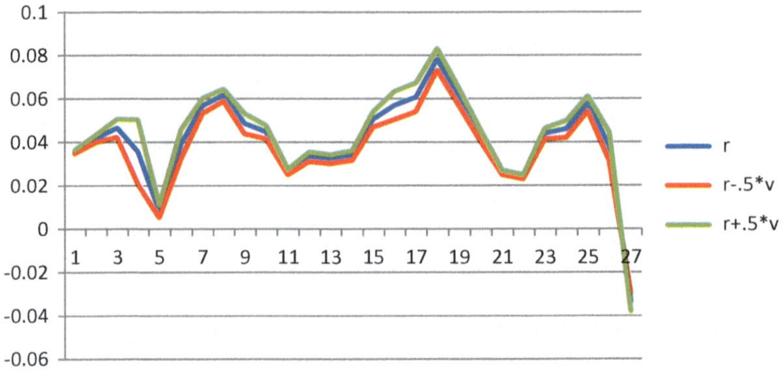

Fig. 18.3 Growth rates and variation coefficients

18.2.2 *Theil's U Coefficient Generalized*

Theil (1961, pp. 31–45) presents a two-dimensional U coefficient as a tool to compare predicted and realized values of a variable; this coefficient can be generalized (U_g) to more dimensions, in which it also varies between 0 and 1. For the case at hand (four regional growth rates), its specification is as follows:

$$U_g = \sum\nolimits_{j>l,i=1,\ldots,4} \left[n^{-1} \sum\nolimits_t \left(r_{it} - r_{jt} \right)^2 \right]^{-0.5} / 2 \sum\nolimits_i \left(n^{-1} \sum\nolimits_t r_{it}^2 \right)^{-0.5}. \quad (18.12)$$

Applied to the quantities of Table 18.4, it produces a value of 0.4315, which is highly significant (Theil 1961, p. 32), confirming the previous findings of severely divergent regional growth rates.

18.2.3 Some Trigonometry

One can measure the angle, α, between the vector of the r_i ($\mathbf{u}_i$) and the r-vector ($\mathbf{u}$) (see Johnson and Kiokemeister 1964, pp. 464 and 471); in the present case, one gets

$$\cos\left[\alpha\right] = r\sum_i r_i/2\left[r\left(\sum_i r_i^2\right)\right]^{0.5}, \tag{18.13}$$

for every observation. Table 18.5 reports the values $a = 1 - \cos[\alpha]$ as the left-hand side of Eq. (18.13) declines with the angle.

The average value in Table 18.5 is 0.1267; the cosine value 0.8733 corresponds roughly to an angle of 29°. But as Table 18.5 shows, this angle is highly variable over the years (in fact, it fluctuates between roughly 6° and 99°), confirming the previous findings.

Table 18.5 Values of *a*

0.006553
0.019692
0.041328
0.806065
1.162188
0.341877
0.014193
0.007627
0.048791
0.043803
0.061842
0.072778
0.070126
0.04326
0.032122
0.079973
0.064643
0.009218
0.016676
0.031448
0.028915
0.042393
0.025454
0.033549
0.021761
0.153892
0.140942

18.2.4 Correlation Analysis

Table 18.6 reports the lower triangle entries of the correlation matrix for r_1 through r_4.

For a complete $r_{ij} = 1$ matrix, all of the characteristic values (i.e., eigenvalues) would be zero, except for one equal to n, the order of the matrix and its trace; for an identity matrix, they all would be 1. In the present case, they are 0.1042, 0.1571, 0.7944 and 2.9442, which is an intermediate case, but dominated by one value, this outcome arising from the presence of three relatively high correlation coefficients. Consequently, the fourth, and accessorily the third, characteristic vectors (i.e., eigenvectors) are really relevant, which has been verified with the characteristic elements of Table 18.6; so the growth rate correlations could be condensed for nearly 75% to λ_4, the eigenvalue, and l_4, the corresponding eigenvector (see values in Table 18.7 in the next subsection).

18.2.5 Synthesis

Table 18.7 summarizes the principal results of this section.

The quantities in Table 18.7 give a fair overview of the characteristics of the regional growth rates analyzed, in particular their spatial spread and its evolution. The correlation between v and a (=1 – cos[α]) is weak (0.1545), although significant at the 0.5 level. This result is due to v being a coefficient of variation, the standard deviation related to the average value, so even if this value were to be stable, the spread of spatial growth rates would increase with increasing overall growth.

Table 18.6 Correlation matrix

	r_1	r_2	r_3	r_4
r_1	1			
r_2	0.464877	1		
r_3	0.330849	0.853499	1	
r_4	0.336988	0.883	0.848667	1

Table 18.7 An overview of results

Values	V	U	α	λ_4, l_4
Average	0.000696		29°10′	
Minimum	−0.000890		6°34′	
Maximum	0.002955		99°60′	2.9442
Observed		0.4315		0.3195
Observed				0.5595
Observed				0.5374
Observed				0.5441

18.3 Chapter Conclusions

Spatial Econometrics (Paelinck and Klaassen 1979) enumerates five proposed principles as guidelines to the correct specification of spatial econometric models: spatial interdependence, spatial asymmetry, allotopy (the distance action of exogenous variables), non-linearity, and—obviously—the presence of a topology.

Recent work, as reported on here, shows that other specific characteristics of spatial econometric models should be highlighted. One of these characteristics is that specification of these models should take into account the fact that groups of regions, not necessarily contiguous (e.g., urban hierarchical levels), could reveal different behavioral patterns, as the preceding results confirm with different techniques; this is one of the aspects of non-convexities, another important feature of theoretical spatial economics and spatial econometrics, to which the next chapter is devoted.

References

Aitchison, J., & Brown, J. (1957). *The lognormal distribution*. Cambridge: Cambridge U. Press.

Ancot, J.-P., & Paelinck, J. (1983a). Interdépendances spatiales et modèles économétriques du développement multirégional. In B. Planque (Ed.), *Le développement décentralisé: dynamique spatiale de l'économie et planification régionale* (pp. 107–116). Paris: LITEC.

Ancot, J.-P., & Paelinck, J. (1983b). The spatial econometrics of the European FLEUR-model. In D. Griffith & A. Lea (Eds.), *Evolving geographical structures* (pp. 229–246). The Hague: Martinus Nijhoff.

Johnson, R., & Kiokemeister, F. (1964). *Calculus with analytic geometry*. Boston: Allyn and Bacon.

Ord, J. (1972). *Families of income distributions* (Griffin, Statistical Monographs and Courses Series, No. 30). London: Hafner Publishing Company.

Paelinck, J., & Klaassen, L. (1979). *Spatial econometrics*. Farnborough: Saxon House.

Tanner, J. (1961). *Patterns affecting the amount of travel* (Technical Paper No. 51). London: Road Research Laboratory, HMSO.

Theil, H. (1961). *Economic forecasts and policy* (2nd ed., revised). Amsterdam: North Holland.

Chapter 19
Traveling with the Salesman...

Quadratic Assignment Problems (QAPs) are well documented in the literature, but certain analogies and properties have not been discussed. This is the objective of this chapter, which addresses three classical problems: the Traveling Salesman Problem (TSP), the Matrix Permutation Problem (MPP), and the Koopmans–Beckmann Problem (KBP). Solutions are discussed as a function of the restrictions (side conditions) associated with them, some of these being derived from symmetry properties of the problems tackled. Examples in the field of spatial economics are given.

QAPs have been amply discussed, especially computer algorithms rendering efficient solutions for them (Paelinck et al. 1983, pp. 251–257 and 273–279).

Some formal similarities—and notably symmetry properties—seem to be underexposed in the literature. This chapter addresses some of these issues and tries to establish whether or not they can be of some help in deriving efficient binary solutions. An empirical example from cluster analysis is given (also see Chap. 6).

19.1 The Traveling Salesman Problem

In brief, the TSP consists of finding the solution to the task, for a salesman, to travel along the shortest path, or route, or itinerary, joining a certain number of places to be visited in succession (for an inverse assignment problem, see Paelinck 2000).

In purely formal terms, like the two other problems to be discussed, it can be formulated as follows:

$$\min_{\mathrm{x}} \mathbf{x}'\mathbf{A}\mathbf{x}, \tag{19.1}$$

D. A. Griffith, J. H. P. Paelinck, *Morphisms for Quantitative Spatial Analysis*, Advanced Studies in Theoretical and Applied Econometrics 51,
https://doi.org/10.1007/978-3-319-72553-6_19

$$\mathbf{Jx} = \mathbf{i}, \tag{19.2}$$
$$\widehat{\mathbf{x}}x = x. \tag{19.3}$$

Expression (19.1) renders the specific condition of the problem through the structure of matrix **A;** in the case of a TSP, this matrix is derived from a distance matrix, **D**. **x** is a binary (0, 1) column vector, in this case the order in which places have to be visited; x_{ij}, e.g., denotes that place *i* has to be visited as the *j*th one in the ranking. The specificity of matrix **A** can best be illustrated by a term from Eq. (19.1), $x_{ij}a_{ij;k,j+1}x_{k,j+1}$, in which $a_{ij;k,j+1}$ represents the travel cost (e.g., distance, time, fare) incurred by visiting place *i* as number *j* and *k* as number *k* + 1.

Conditions (19.2), where **i** is the unit column vector, known as the assignment conditions, state that each place in the ranking has to be occupied once and only once; if **x** is of order n^2x1, **J**, a binary matrix, is of order $2nxn^2$. However, because one of the equations is not independent of the others—for obvious reasons—it should be reduced to order $(2n - 1)n^2$. Finally, Eq. (19.3) state the binarity (0, 1) conditions ($\widehat{\mathbf{x}}$ is a diagonal matrix).

A first reduction in the size of the problem can be obtained by substituting Eq. (19.2) into Eq. (19.1); this substitution leads to the reduced objective function

$$\min_{\mathbf{x}^*} \mathbf{x}^{*'}\mathbf{B}\mathbf{x}^* + \mathbf{b}'\mathbf{x}^*, \tag{19.4}$$

in which the constant term is neglected, and where **x*** is of order $(n - 1)^2$-*by-1*, because $(n - 1)^2 = n^2 - 2n + 1$. But additional—weak—constraints can be introduced now, to wit the weakened assignment conditions become

$$\mathbf{J}^*\mathbf{x}^* \le \mathbf{i}^*, \tag{19.5}$$

which is due to the fact that—after the problem has been solved—certain routes can be deduced from the equation

$$x^{**} = J^{**^{-1}}(-J^*x^* + i^*), \tag{19.6}$$

in which **x**** are the variables excluded from Eq. (19.4), **J*** and **J**** being the composing parts of matrix **J** that has been conformably partitioned. A handy way to do this derivation—guaranteeing that **J**** is regular—is to take for **x**** a full set of rankings for one place—their number being *n*—and for all other places a given ranking (e.g., the last, *n*-th, one); the remaining values belong to **x***.

Under these conditions $n - 2 \le \mathbf{i'x^*} \le n - 1$. Indeed, if the full preceding set has its last value as the optimal one, then this value no longer can be obtained from the remaining places, and they have to find their unit values in **x***. In the reverse case, two unit values appear in **x****, so only *n* – 2 remain for **x***. We also study here a weakening of condition (19.3) to

$$0 \leq x^* \leq i, \tag{19.7}$$

in order to find out how strong the following symmetry property is in obtaining binary solutions.

Indeed, the same values are found for the objective function if the trip is made in the reverse order! Therefore, a vector **y***, symmetric to **x***, satisfies

$$x^{*'}Bx^* + b'x^* = \varphi^* = \psi^* = y^{*'}By^* + b'y^*. \tag{19.8}$$

Conditions (19.5), together with the symmetry conditions, except for y_{i2}, apply to vector **y***. The reason is that place 1, the starting point of a trip, is fixed, so only places 2 through n remain. Now, place 2 can be eliminated by Eq. (19.6) and rank n for all places 3,...,n also is eliminated. Thus, for place 3, e.g.,

$$x_{32}^* = 1 - c^{*'}x_3^* \tag{19.9}$$

and

$$y_{33}^* = x_{3,n-1} \tag{19.10}$$

$$\cdots$$

$$y_{3,n-1}^* = x_{33}^*. \tag{19.11}$$

However, this result does not guarantee that

$$\mathbf{y}^{*'}\mathbf{x}^* = 0, \tag{19.12}$$

(namely, when nonbinary values are present). Consequently, this condition has to be studied, too; it plays an important part in the solution of interest.

The preceding specification has been applied to a *5-by-5* case, for which matrix **D** is presented in Table 19.1.

Under the preceding side conditions—except for Eq. (19.12)—the solutions are those appearing in Table 19.2, with $\varphi^* = \psi^* = -0.5417$.

Not only do those conditions not suffice to render binary results, but side conditions (19.12) are not satisfied, either. If these side conditions are introduced, the results are those appearing in Table 19.3, with $\varphi^* = \psi^* = 0.21875$.

Again, the side conditions do not suffice to render binary results; the binary vector, which appears in Table 19.4, could be obtained only by adding the quadratic condition

$$\mathbf{x}^{*'}\mathbf{x}^* = \mathbf{i}'\mathbf{x}^*. \tag{19.13}$$

For this solution, $\varphi^* = \psi^* = 1$, the total travel cost being 22. The discrepancy between 1 and 22 is due to the omission of the constant in Eq. (19.4), obviously equal to 21. Therefore, the **x**-order is $[x_{42}, x_{23}, x_{34}, x_{55}]$, with the **y**-order being $[y_{52}, y_{33}, y_{24}, y_{45}]$.

Table 19.1 Matrix **D**

Places	1	2	3	4	5
1	0	2	4	3	6
2	2	0	1	4	7
3	4	1	0	5	8
4	3	4	5	0	9
5	6	7	8	9	0

Table 19.2 First TSP results

Indices	x^*	y^*
32	0	0.4167
33	0.0833	0.5000
34	0.5000	0.0833
42	0.5833	0
43	0.4167	0
44	0	0.1167
52	0.4167	0.5833
53	0	0
54	0	0

Table 19.3 Second TSP results

Indices	x^*	y^*
32	0.3125	0
33	0	0.6875
34	0.6875	0
42	0.6875	0
43	0.3125	0
44	0	0.3125
52	0	1
53	0	0
54	0	0

Table 19.4 Binary TSP results

Indices	x^*	y^*
32	0	0
33	0	1
34	1	0
42	1	0
43	0	0
44	0	0
52	0	1
53	0	0
54	0	0

Table 19.5 Unconstrained TSP results

Indices	x^*
32	0.3125
33	0.2143
34	0.1875
42	0.6875
43	0.3125
44	0
52	0
53	0
54	0.2875

Comparing Tables 19.3 and 19.4, one can see that if the quantities in Table 19.3 are rounded off, one obtains Table 19.4. Whether or not this rounding correspondence is true for larger problems has to be investigated, probably by Monte Carlo methods, especially because nothing is revealed about the second-order conditions. For the exercises of this chapter, a reduced gradient method was used (Fylstrom et al. 1998).

The same cannot be said for Table 19.2, where x_{52} even has a positive value, and, of course, if conditions (19.12) and (19.13) are skipped, the result is even more fuzzy, as Table 19.5 shows; the φ^*-value is, to be sure, the lowest one obtained (-1.4241).

19.2 The Matrix Permutation Problem

Formally, this MPP problem also is specified by Eqs. (19.1) through (19.3); only the contents of matrix **A** are different; the problem consists of finding that permutation of rows and columns of a given matrix that would maximize (or minimize) the sum of the cells above (or below) the main diagonal. A classic application is the (quasi)-triangulation of a (regional) input–output matrix.

Here, the max-sup problem is treated and again the impact of restrictions (19.9) through (19.13) studied. In doing so, four more remarks merit being stated.

The first one is that the **D**-matrix of Table 19.1 cannot be used any more, because all *n!* permutations would be optimal, the symmetry of the matrix implying that the sup- and sub-sums are equal. Table 19.6 is the result from Table 19.1 with a one-cell shift, to the right for the uneven rows, and to the left for the even ones, with the last term of the last row becoming the first nonzero one of row one.

Second, unlike in the TSP, place one is no longer given, but must be determined optimally.

Third, now Eq. (19.12) renders the value 1, because the middle row-and-column is kept in place during the symmetry permutation, the reason being the unevenness of the relevant number of rows and columns.

Finally, one has $\psi = \sigma - \varphi$, where σ is the sum of all terms of the matrix.

Table 19.6 Shifted D-matrix

Places	1	2	3	4	5
1	0	9	2	4	3
2	6	0	2	1	4
3	7	4	0	1	5
4	8	3	4	0	5
5	9	6	7	8	0

Astonishingly enough, the unconstrained solution appears to be binary, the optimal reshuffling being [x_{41}, x_{32}, x_{53}, x_{14}, x_{25}], and $\varphi^* = -12$, corresponding to an upper sum of 60 and a lower one of 38 (so $\psi^* = -34$). This outcome is all the more unexpected, because the preceding TSP is only of complexity $(n - 1)!$, whereas the MPP is of complexity $n!$. In contrast, one reason for this outcome could be the non-symmetry of the shifted matrix **D**, upon which more reflection is necessary. Otherwise, this outcome might as well be a matter of chance, and, again, solutions need to be computed for large matrices and then studied.

19.3 The Koopmans–Beckmann Problem

The KBP (Koopmans and Beckmann 1957) is a locational assignment problem, addressing the issue of locating establishments separately in given sites so as to minimize total transportation costs; this formulation means that matrix **D** of Table 19.1 again can be used.

A difference exists between the KBP and the TSP, in the sense that distances are now to be multiplied by the volume of goods shipped and the unit transportation costs in order to obtain matrix **A** of Eq. (19.1). To be fully clear, index i now stands for an establishment and j for a location. In the exercise to follow, we assume the same number for both (i.e., 5); furthermore, we suppose the *goods times unit transportation costs* to be equal to the corresponding terms of the shifted **D**-matrix (Table 19.6), but this assumption is invoked just to set up the exercise. It gives rise to $(n - 1)^2 = 16$ transport cost matrices, which are not reproduced here.

Again, the unconstrained solution happens to be binary, with an optimal assignment [x_{31}, x_{52}, x_{43}, x_{24}, x_{15}], giving rise to a (reduced) cost of $\varphi^* = -939$, corresponding to a total cost of $\varphi = 464$. The program pick lower transportation costs, the lowest being *4* and the highest being 81; the resulting transport costs matrix **T** is reproduced in Table 19.7, the entries corresponding to index j, the locations.

One remark about the constraints still merits stating. If one compares the KBP to the MPP, one should notice that the symmetry value ψ^* cannot be obtained by subtraction from a known value σ; however, what could be done is to introduce this value as an endogenous variable, to be computed together with the binary ones. For the preceding problem, these values, respectively, are −930 and −1869, showing that the total transportation cost would be 9 units higher.

Table 19.7 Transport costs matrix for the KBP

Sites	1	2	3	4	5
1	0	10	4	12	12
2	14	0	8	24	63
3	16	5	0	15	64
4	6	16	5	0	54
5	81	12	32	21	0

19.4 Dynamic Cluster Analysis

Activity complexes have to be developed over time, so some dynamics should be introduced. Once more quadratic assignment can be a suitable method for solving the dynamic industrial complex problem (DICP; also see Paelinck and Mur 2013).

Suppose that three activities are to start over a given time horizon; the inter-temporal character of the problem derives from the fact that investment funds might be limited inside each time period. The idea then is to exploit the following facts:

- Prices rise over time, a well-known fact for important projects (the Chunnel, or Channel Tunnel in Western Europe is a much quoted example).
- The presence of activities on a site generates externalities.

Consider the definitions:

- x_{it} is a binary variable, meaning that activity i is to be started in time period t.
- a_{it} is the cost of implementing activity i in period t.
- b_{ijt} is the percentage/100 savings accruing to activity i due to the presence in time period t of activity j.

The three activity-three time period program can then be specified as follows:

$$\mathrm{Min}(\varphi_1 + \varphi_2 + \varphi_3) \tag{19.14}$$

s.t.:

$$\begin{aligned}\varphi_1 \triangleq\, & a_{11}x_{11} + a_{12}x_{12}(1 - b_{122}x_{21} - b_{132}x_{31}) + a_{13}x_{13}\left[1 - b_{123}(x_{21} + x_{22})\right. \\ & \left. - b_{133}(x_{31} + x_{32})\right],\end{aligned} \tag{19.15}$$

$$\begin{aligned}\varphi_2 \triangleq\, & a_{21}x_{21} + a_{22}x_{22}(1 - b_{212}x_{11} - b_{232}x_{31}) + a_{23}x_{23}\left[1 - b_{213}(x_{11} + x_{21})\right. \\ & \left. - b_{233}(x_{31} + x_{32})\right],\end{aligned} \tag{19.16}$$

$$\begin{aligned}\varphi_3 \triangleq\, & a_{31}x_{31} + a_{32}x_{32}(1 - b_{312}x_{11} - b_{322}x_{21}) + a_{33}x_{33}\left[1 - b_{313}(x_{11} + x_{12})\right. \\ & \left. - b_{323}(x_{21} + x_{22})\right],\end{aligned} \tag{19.17}$$

$$\mathbf{Jx} = \mathbf{i} \tag{19.18}$$

Table 19.8 Data for problem (19.14)–(19.18)

Parameters	Values
a_{11}	1
a_{12}	2
a_{13}	3
a_{21}	2
a_{22}	3
a_{23}	4
a_{31}	3
a_{32}	4
a_{33}	5
b_{122}, b_{123}	0.3
b_{132}, b_{133}	0.2
b_{212}, b_{213}	0.2
b_{232}, b_{233}	0.4
b_{312}, b_{313}	0.4
b_{322}, b_{323}	0.1

If one writes out matrix **A**, one can see that its structure resembles that of the MPP; this fact is due to the presence of externalities, in the form of plants that should exist before another plant is started. This problem feature explains why the square sub-matrices of **A** have a lower block-triangular structure.

Table 19.8 presents the data used to test this specification.

Solving this problem by a linearization method for QAPs (see Sect. 16.2.1.2), one obtains the optimal x_{it}-vector $[x_{13}, x_{21}, x_{32}]$ with a value of 7.1 for the objective function (19.14).

Isard et al. (1959; see there Sect. 1.3, especially Tables pp. 54–55, 68 and 98–99) make real-world data available; these data have been used in the following exercise. The petrochemical complex chosen was the chain ethylene–ethylene glycol-dacron polymer-dacron staple, the oil refinery (index 0) being already present, so four activities have to be planned over time. Table 19.9 presents the data; some of these data were not available, and hence had to be hypothesized, stars marking them. The symbols appearing in this table have the same meaning as before, it being understood that there are now four periods. The first three columns include the original data; only columns four and five have been included in the calculations. Finally, an inflation rate of 10% has been postulated.

The result is x_{11}, x_{22}, x_{33}, x_{44}, the technical sequence of the petrochemical process, with an objective function value $\varphi = 30.39$. However, in a different simulation, the first two activities were inverted, which amounted to activity 2 provisionally buying ethylene from outside the complex, while activity 1 was under construction.

Many refinements could be introduced. One possibility is discounting, though the program presented computes the total expected optimal cost for the complex; i.e., the financial envelope to be provided for. Another possibility is the introduction

Table 19.9 Data for a petrochemical complex

Coefficient	Unit inventory	Activity level	Scale coefficient	Total inventory	Externalities
a_{11}	2.65	25	0.67	22.90	–
a_{22}	4.31	1	0.625	4.31	–
a_{33}	3	1*	0.67*	2.16*	–
a_{44}	3	1*	0.67*	2.16*	–
b_{011}, b_{012}, b_{013}, b_{014}	–	–	–	–	0.1166
b_{122},, b_{123}, b_{124}	–	–	–	–	0.0275
b_{212}, b_{213}, b_{214}	–	–	–	–	0.0070
b_{232}, b_{233}, b_{234}	–	–	–	–	0.0054
b_{322}, b_{323}, b_{324}	–	–	–	–	0.0037
b_{342}, b_{343}, b_{344}	–	–	–	–	0.0037
b_{432}, b_{433}, b_{434}	–	–	–	–	0.0059

of financing constraints, possibly as a function of the duration of each of the subprojects.

The technique exposed is sufficiently flexible to accommodate these two suggested refinements and also other ones that could creep up in special programs.

19.5 Chapter Conclusions

In the first place, the present study has tried to clarify the structures of the core matrix—**A**—of QAPs; four fundamental ones appear to be present, to wit the TSP, the MPP, the KBP, and the DICP.

Moreover, testing them with a small sample suggests that a difference in structure could lead to different solution procedures. The most complex one includes the insertion of the assignment conditions into the objective function, the simultaneous use of the symmetric (dual) problem, the orthogonality of the primal and dual variables, and finally a simple binarity condition. The TSP needs all of these features, the other ones requiring only the insertion procedure. The symmetry of the TSP core matrix is conjectured to be responsible for the nature of the solution, although further tests—with much larger samples—are needed to find this out.

References

Fylstrom, D., Lasdon, L., Watson, J., & Waren, A. (1998). Design and use of the Microsoft Excel solver. *Interfaces, 28*(5), 29–55.

Isard, W., Schooler, E., & Vietorisz, T. (1959). *Industrial complex analysis and regional development: A case study of refinery-petrochemical synthetic fiber complexes and Puerto Rico.* Cambridge, MA: MIT Press.

Koopmans, T., & Beckmann, M. (1957). Assignment problems and the location of economic activities. *Econometrica, 25*, 53–76.

Paelinck, J. (2000). *Controlling complexity in spatial modelling* (Working Paper No. 362). Munich, Germany: CESifo Group.

Paelinck, J., & Mur, J. (2013). Theoretical spatial economics and spatial econometrics: Time-and-space non-convexities galore. *Estadistica Espñola, 55*(180), 13–32.

Paelinck, J., Ancot, J.-P., & Kuiper, J. (1983). *Formal spatial economic analysis.* Aldershot: Gower.

Chapter 20
Complexer and Complexer, Said Alice …

Because the notion of a corp allows even more of a systematization than appears in Sect. 16.1.3, this chapter begins with that concept, giving some examples of its application. For the next sections, discussions in the early 2000s with Roger Stough, and assistance from Rajendra Kulkarni and Guang Yang, all active then with the School of Public Policy, George Mason University, are gratefully acknowledged.

The formal treatment of the problems justifies their presence in this morphism series of analyses.

20.1 Corps Anew

Some examples of corps are mentioned at the end of Sect. 16.1.3; however, other results are available. One of those results (more results appear in Paelinck 2000) is for a 10-by-10 input–output matrix for the United Kingdom in 1950 (Stone and Croft-Murray 1959, p. 33); it runs as follows.

Only the first three iterations produce at least one corps; the very first round produces a complete set of three corps. These rounds together account for, respectively, 40.60%, 18.34%, and 11.62% of all transactions in the table; the following are the generated corps:

Round 1:

(a) Agriculture, forestry, fishing ⇒ food, drink, tobacco ⇒ agriculture, forestry, fishing.
(b) Mining, quarrying ⇒ gas, electricity, water ⇒ chemicals, allied trades ⇒ textiles, leather, clothing ⇒ other manufacturing ⇒ building, contracting ⇒ mining, quarrying.
(c) Metals, engineering, vehicles ⇒ other production, trade ⇒ metals, engineering, vehicles.

D. A. Griffith, J. H. P. Paelinck, *Morphisms for Quantitative Spatial Analysis*, Advanced Studies in Theoretical and Applied Econometrics 51,
https://doi.org/10.1007/978-3-319-72553-6_20

Round 2:

Agriculture, forestry, fishing ⇒ textiles, leather, clothing ⇒ mining, quarrying ⇒ chemicals, allied trades ⇒ agriculture, forestry, fishing.

Round 3:

Mining, quarrying ⇒ textiles, leather, clothing ⇒ chemicals, allied trades ⇒ metals, engineering, vehicles ⇒ mining, quarrying.

Round 1: (a) reproduces the agricultural food cycle, (b) focuses on chemical production, whereas (c) depicts the metalworking complex. Rounds 2 and 3 refer to more involved technologies, but the level of aggregation is too coarse to allow disentangling them readily. Nevertheless, the aforementioned filières are clearly present; this would not necessarily be the case for a regional input–output table describing the structure of a regional economy, especially a small one, due to the absence of certain activities and/or the presence of important competitive imports.

Very few, if any, spatial elements are present in the preceding exercise. Certainly, the sectors analyzed are located somewhere in geographical space, and as a function of the definition of the input–output tables used, coefficients can be dependent on competitive imports, but spatiality stops there. Other corps—organizational ones, called clubs, to mention only one example of these—also could be identified.

A new typology finally emerges from the foregoing analyses; Fig. 20.1 portrays it, classifying the various types of clusters mentioned earlier. Two new notions need to be introduced, to wit clans, transposing complexes defined in Sect. 16.1.2, and resting on technical relations, to contact coefficients, and clubs, generalizing the idea of corps, also defined earlier, and referring to the same technical relations and the same contact coefficients.

Figure 20.1 also indicates that *mixed* local concentrations can occur (see the arrows). A research program for the future should include identification of the relative strength of the locational forces at work, following the principles of spatial econometrics (Paelinck and Klaassen 1979). Such an analysis is indispensable to the devising of efficient (multi)regional policies.

A possible approach to solving this problem is by means of nonplanar graphs (for planar graphs, see Tutte 2001, pp. 285–326; Diestel 2000, pp. 67–94). Multiple matrix relations can be visualized as a graph, mostly nonplanar, a planar graph being characterized by the fact that a particular location of its nodes in two-dimensional space would not allow its edges to cross at points otherwise than nodes of the graph. One way of jointly using two corps results is to count the number of joint nodes, maximizing their sum. Table 20.1 presents a hypothetical matrix of two types of bi-directional relations; the second quantity in a cell relates to the second type of relation, say a club relation versus a corps relation (an absence of the first relation is designated by a star).

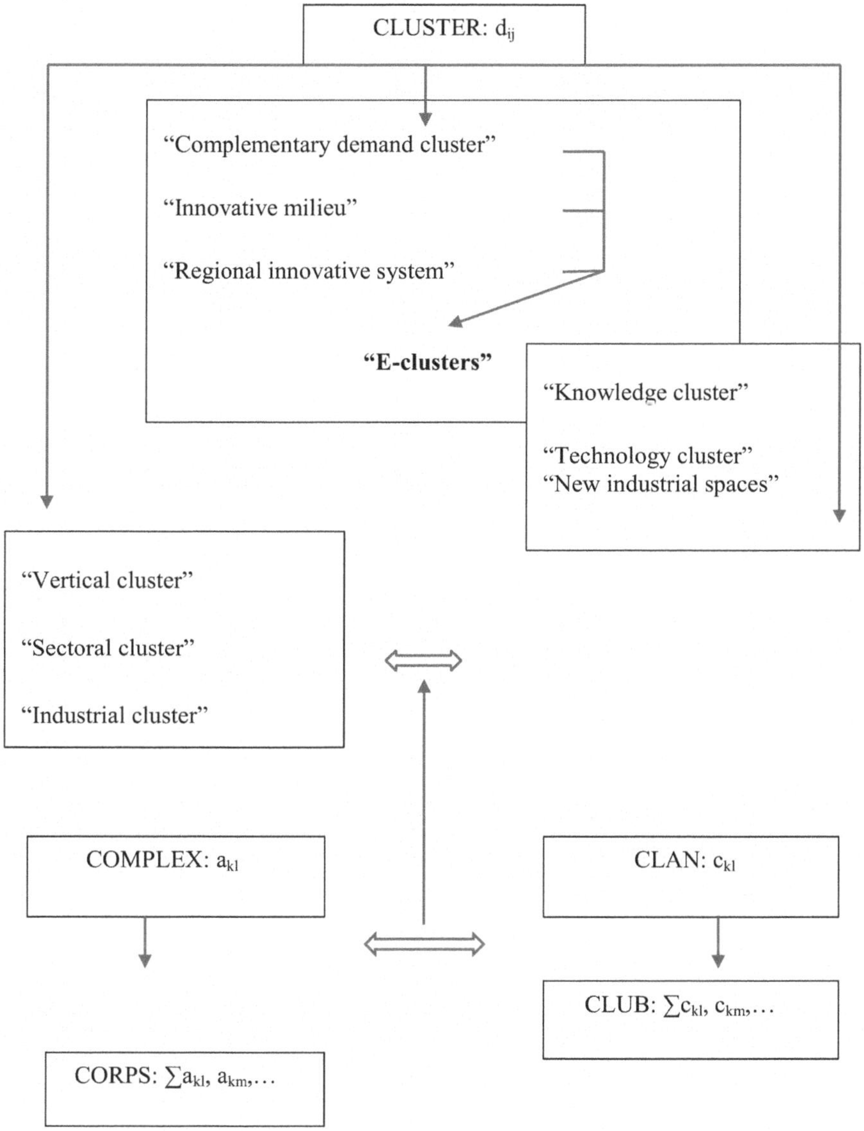

Fig. 20.1 A conceptual flow diagram

If one numbers relations of the first type by 1 and 2, and those of the second type by A and B, we get the following results for the common nodes: $1A = 2$, $1B = 2$, $2A = 2$, $2B = 4$; thus, the grouping of both relations would be $1A/2B$.

If one now wants to integrate the bi-dimensional intensities as shown in Table 20.1, one has to formulate the following mathematical program:

Table 20.1 Corps and club relations

U	1	2	3	4	5	6	7	8	9	10	11	12	13	14	15	16
1		0.1					0.2									
2	0.2		0.3													0.3
3		0.4		0.4											0.2	
4			0.1		0.6			0.2				0.1				
5				0.3		0.01										
6					0.4		0.2									
7						0.5		0.1								
8	0.3			0.3			0.3			0.2						
9										0.3						0.1
10								0.5	0.1		0.5			*0.3		
11										0.2		0.2;1		*0.1		
12				0.2							0.3;0.1		0.2			
13												0.4		0.6		
14										0.2	0.2		0.1		0.9	
15			0.4											0.3		0.2;0.3
16		0.4							0.3						0.3;0.1	

$$\begin{aligned}\text{Max} : \varphi = \; & \lambda\textstyle\sum_{1,ij}\left[(a_{ij}+b_{ij})x_{ij}x_{ij}^{*}\right] + (1-\lambda)\textstyle\sum_{2,ij}\left[(a_{ij}+b_{ij})x_{ij}x_{ij}^{*}\right] + (1-\lambda) \\ & \times\textstyle\sum_{3,ij}\left[(a_{ij}+b_{ij})x_{ij}x_{ij}^{*}\right] + \lambda\textstyle\sum_{4,ij}\left[(a_{ij}+b_{ij})x_{ij}x_{ij}^{*}\right]\end{aligned} \tag{20.1}$$

s.t.:

$$\lambda = \lambda^2, \tag{20.2}$$

where the sums run over the four cases of common nodes previously distinguished, and the $x_{ij}s$ and $x_{ij}^{*}s$ are binary variables, corresponding to the presence or absence of one and/or the other of the relational links. The result is $\lambda = 1$, again giving the grouping $1A/2B$, with $\varphi = 1.7$.

This method can be easily generalized to more than two circuits and more than two relations.

20.2 A Topography of Complexes

As joint economic-spatial entities, complexes should be considered from both points of view. What follows addresses the topographical one.

20.2.1 Circumscribing Clusters

The first exposition is about the method, after which some supporting mathematics are presented.

20.2.1.1 The Method

Proper topographical analysis of industrial clusters was brought to the attention of the author by Stough, Button, Kulkarni, and Yang, as previously mentioned. A remarkable result of their investigation was the discovery of a significant relation between interindustry links and geographical proximity of production units in urban areas, this relation in turn being a function of the local geography—in particular, the existing transportation network—and the history of the cluster. The spatial analysts mentioned previously presents a topographical analysis of 33 sectors belonging to the Information and Communication activities in the Austin (A), Boston (B), and Washington, DC (W) metropolitan regions. The analysis discussed hereafter concerns the envelopment of the centers of gravity of the sectors concerned, the method consisting of deriving a convex quadratic envelope for those points, and the specification of the model being as follows.

For the non-weighted center of gravity of the points mentioned (coordinates x^*, y^*), for each activity i one computes the quantities $u_i = (x_i - x^*)^2$, $v_i = (y_i - y^*)^2$, $z_i = (x_i - x^*)(y_i - y^*)$. Let **u, v, z,** be the column-vectors of those magnitudes, and $\mathbf{X} = [\mathbf{u,v,z}]$ be the resulting matrix. One then builds a 3-by-3 matrix $\mathbf{X'X}$; defines a column-vector **e** of three coefficients a, b, c; and derives the following mathematical program

$$\text{Min}: \varphi = \mathbf{e}'\mathbf{X}'\mathbf{X}\mathbf{e} - 2\mathbf{i}'\mathbf{X}\mathbf{e}, \tag{20.3}$$

s.t.

$$a\mathbf{u} + b\mathbf{v} + c\mathbf{z} - \mathbf{i} \leq \mathbf{0}, \tag{20.4}$$

One knows that the quantities resulting from the analytic geometry used are all symmetric (ellipses or, their limit cases, circles; hyperbolas, parabolas). The central idea then is to eliminate that symmetry by means of a logarithmic transformation. After looking for an excellence criterion for the two specifications (natural and logarithmic), and in the absence of a classical statistical criterion (an R^2, for example), one may opt for the area delimited by the computed curves and more precisely in natural values (i.e., in anti-logarithms for the logarithmic transformation). Table 20.2 presents the coefficients computed from the logarithmic data using mathematical program (20.3) and (20.4).

From these tabulated results, corrected for orientation, the following natural values (anti-logarithms) can be derived for the local areas (original quantities multiplied by 100): A(3.6748), B(53.7348), W(11.3409). Of note is that the Boston area reveals an ellipsoid result (in natural numbers, it is hyperbolic), but, as is logical, with a relatively large area.

Table 20.3 summarizes the results in terms of areas (again original quantities times 100).

These results demonstrate that the logarithmic calibration does not necessarily reduce the enveloping areas. In all cases, the ellipsoidals exceed the corresponding rectangular ones. Such a result is possible if the observations are concentrated near the corners of a rectangle; a simple case is that of a square in which each corner hosts one observation, the computed circle then being circumscribed.

Figures 20.2, 20.3 and 20.4 depict all of the treated cases.

20.2.1.2 Supporting Mathematics

This section summarizes the mathematical foundation for the preceding analyses.

Table 20.2 Logarithmic ellipsoid analysis

City	A	B	W
A	199.1778	12.7984	7.6980
B	35.4853	35.4528	114.5175
C	−12.9154	−40.9552	21.8720

Table 20.3 Areas

Area	A	B	W
Natural	2.5480	∞	11.6600
Logarithmic	3.6748	53.7348	11.3409
Rectangular	2.3819	23.9201	8.0658

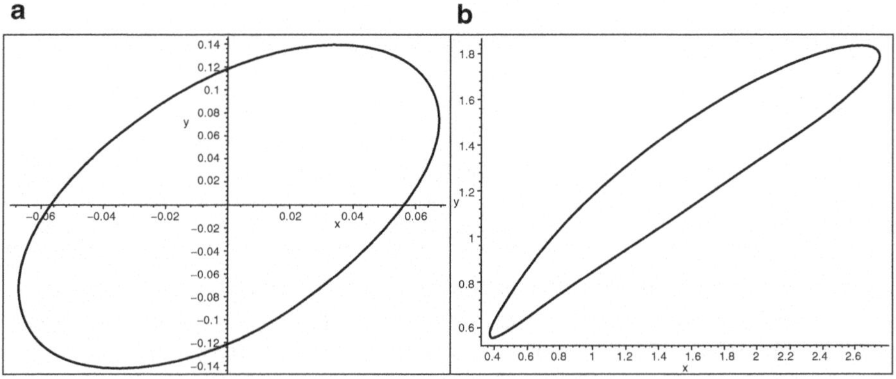

Fig. 20.2 Austin. *Left* (**a**): Original envelopment. *Right* (**b**): Log-transform

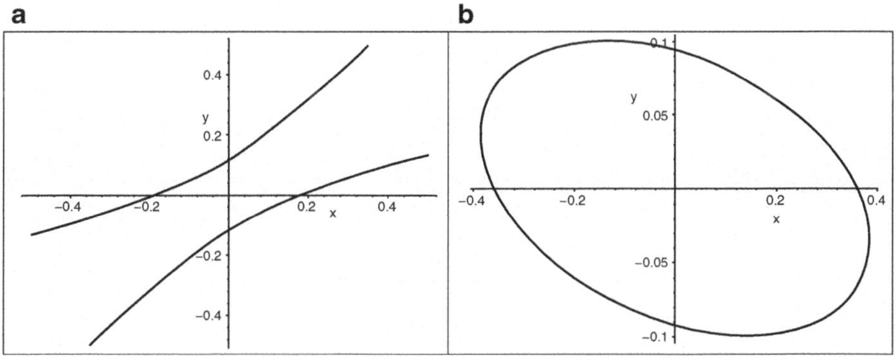

Fig. 20.3 Boston. *Left* (**a**): Original envelopment. *Right* (**b**): Log-transform

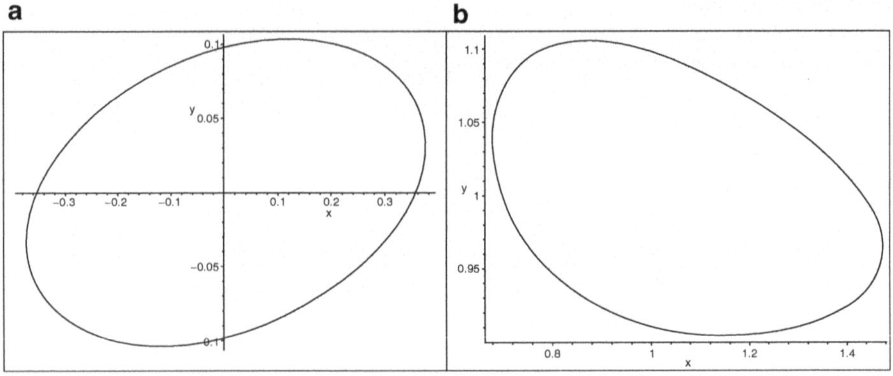

Fig. 20.4 Washington, DC. *Left* (**a**): Original envelopment. *Right* (**b**): Log-transform

Elliptic Area

The ellipse has as the following characteristic equation:

$$ax^2 + by^2 = \rho^2, \tag{20.5}$$

where x and y are Cartesian coordinates, ρ is a given constant, and a and b are inverses of the squares of the semi-axes; these latter two quantities are here strictly positive parameters, although other cases are possible (e.g., hyperbolas and parabolas). Equation (20.5) is written as follows in polar coordinates:

$$r^2\left(a\ \sin^2\varphi + b\ \cos^2\varphi\right) = \rho^2, \tag{20.6}$$

where now r is the radius, and φ *is* the angle expressed in radians.

From Eq. (20.6), the elliptic area is computed as

$$s = \int_0^{2\pi} r \int_0^{\rho(a\ \cos 2\varphi + b\ \sin 2\varphi)} rdrdp\ d\varphi, \tag{20.7a}$$

$$= 1/2\rho^2 \int_0^{2\pi} \varphi\left(a \cos^2\varphi + b \sin^2\varphi\right)^{-1} d\varphi. \tag{20.7b}$$

The value of the definite integral (20.7b) is known (CRC 1959, p. 313, No 419) and equal to

$$s = \pi\rho^2(ab)^{-1/2}, \tag{20.8}$$

The Reduction of an Oblique Quadratic to a Canonical Form

Starting with

$$ax^2 + by^2 + cxy = r^2, \tag{20.9}$$

one can apply the following transformation (Fisher and Ziebur 1967, p. 421, Theorem 95-1):

$$a^* = a \cos^2\alpha + b \sin^2\alpha + c \sin\alpha \cos\alpha, \tag{20.10a}$$
$$b^* = a \sin^2\alpha + b \cos^2\alpha - c \sin\alpha \cos\alpha, \tag{20.10b}$$
$$c^* = (b - a) \sin 2\alpha + c \cos 2\alpha = 0, \tag{20.10c}$$

which implies that

$$\cot g 2\alpha = (a - b)/c, \quad 0 < \alpha < \pi/2. \tag{20.10d}$$

Under certain conditions (e.g., the sign of the discriminant or negativity of a^* or b^*), Eq. (20.9) represents a hyperbola, as is the case with Boston (Sect. 20.2.1.1).

Anti-logarithmic Areas

The anti-logarithmic integral of an ellipse not being analytically computable and quadratic approximations having shown themselves fallacious, the following solution has been adopted: position the natural data in such a way that the natural and logarithmic circumscribed rectangles have the same area. The derivation is as follows for coordinate x, that for y being identical.

Define

$$x_{\max} - x_{\min} \triangleq d(x) \tag{20.11}$$

and impose

$$\ln(x_{\max} + h) - \ln(x_{\min} + h) = d(x). \tag{20.12}$$

Then one can compute

$$h = (e^* - 1)^{-1}(x_{\max} - e^* x_{\min}), \tag{20.13}$$

where

$$e^* \triangleq \exp[d(x)]. \tag{20.14}$$

Of note is that a logarithmic projection depends on the absolute values of the natural data; a possibility to be explored is to position the latter in such a way that the logarithmic area is minimal, still respecting condition (20.12).

A Box-Cox Power Transformation

If one replaces variables—necessarily positive—by their Box-Cox transforms (see Reinaud et al. 1995; Griffith et al. 1998), and then extends the expressions x^α and y^β by linear terms, one obtains

$$ax_i^{2\alpha} + by_i^{2\beta} + cx_i^\alpha y_i^\beta + dx^\alpha + ey^\beta = 1. \tag{20.15}$$

For this equation, the elasticity is

$$E_{yx} = \partial y/\partial x = \alpha x^\alpha \left(2ax^\alpha x^\beta + cy^\beta + d\right)\left[\beta y^\beta \left(2by^\beta + cx^\alpha + e\right)\right]^{-1}. \tag{20.16}$$

The empirical elasticities being computable—they may be inferred from the piecewise linear convex hull of the observed points—an iterative procedure, starting with $\alpha = \beta = 1$ and values of a, b, c, d, and e derived from program (20.15) and (20.16), should allow the computing of estimators for α and β (see the references quoted).

The comparison of Box-Cox areas with those obtained from natural variables poses new problems; numerical integration will probably be in order.

20.2.2 Positioning Plants

The following exercise treats the reverse problem, to wit the explicit derivation of relative locations of production units, given their respective distances. The illustrative case is limited here—provisionally—to one production unit per sector. Again beginning setting is with the relationship between input–output relations and plant distances, mentioned previously. Let α_{ij} be the sum of the input coefficients a_{ij} and a_{ji} of the two sectors i and j; then an implicit geographical distance can be defined as

$$d_{ij} \triangleq 2 - \alpha_{ij}. \tag{20.17}$$

This definition satisfies the metric triangularity condition, which enables construction of a distance matrix **D**. One solution to the problem of deriving implicit coordinates is known for Euclidean distances (Tobler 1977). But can a solution be found for Manhattan distances, characteristic for the metric structure of urban areas?

This method reduces to the solution of an overdetermined system of equations between absolute values of the variables concerned; indeed, the Manhattan distance (Minkowski distance for $p = 1$) is defined as

$$d_{ij} = \mid x_i - x_j \mid + \mid y_i - y_j \mid . \tag{20.18}$$

Table 20.4 Results for $p = 1, 2, 3$

Coordinates	Starting values	$p = 1$ & binary variables	$p = 1$ & Eq. (20.17)	$p = 2$	$p = 3$
x_2	2	1.3333	1.7278	−2	1
x_3	3	2.3333	2.7278	−3	2
x_4	1	3.3333	0.7278	−1	3
x_5	4	4	4	−4	4
y_2	1	1.6667	1.2722	1	2
y_3	2	2.6667	2.2722	2	3
y_4	3	0.6667	3.2722	3	1
y_5	4	4	4	4	4

This problem has been solved by mathematical programming. A priori, one should expect multiple solutions; moreover, given the definition of Manhattan distance, symmetrical solutions always exist (Bunch 1989; Nicolle 1965; Rosen 1983) for which the symmetry axes are the coordinate axes and the bisectors. For a 4-by-4 case, the average distance between starting point and solutions is 2.33; for a 5-by-5 case, it decreases to 0.5.

The preceding results directly invite more general applications, in particular the whole class of Minkowski metrics. The method has been applied to the cases of $p = 2$ and $p = 3$; in this latter case, and in all cases with uneven or fractional p, absolute values have to be taken. With respect to the original method, which uses binary variables, a change has been introduced, to wit the use of absolute values.

Table 20.4 gives the results in the 5-by-5 case, referred to earlier, for $p = 1, 2, 3$. Point p_1 was given coordinates (0, 0); the method used was again a reduced gradient method (Fylstrom et al. 1998). The two $p = 1$ cases illustrate the possibility of multiple solutions, independent of symmetry. For $p = 2$, the solution is symmetric to the starting points with respect to the y-axis. For $p = 3$, the coordinates have been interchanged, a symmetry with respect to the bisector. Figure 20.5 illustrates these preceding findings.

The exercises presented in this section should be generalized to larger **D** matrices and to activity clusters with more than one plant per sector, and in each case the solution should be duly qualified (unique or not, independent or symmetric).

The preceding examples imply that a vast field of problems still exists for which spatial econometricians and mathematicians or statistical geographers should be interested in meeting each other (as an example, Getis and Paelinck 2004). The ensuing dialogue from these meetings should encounter no difficulty, because the field mentioned is defined in pre-geographical space (Beguin and Thisse 1979), an ideal meeting place for practitioners of one or the other of these disciplines.

Regional science, by nature interdisciplinary (Paelinck 1982), clearly invites co-disciplinarity.

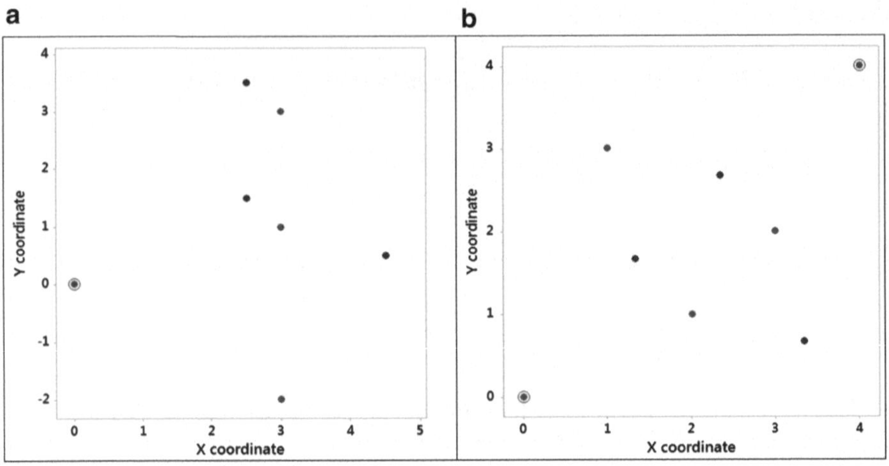

Fig. 20.5 Comparisons of observed (*blue*) and computed (*red*) coordinates. *Left* (**a**): $p = 2$. *Right* (**b**): $p = 3$

20.3 Metropolitan Complexes

The Greater Washington, DC metropolitan region has been the object of a metropolitan complex analysis. Hereafter, the statistical sources are detailed, after which some results—using appropriate techniques—are presented.

20.3.1 Statistical Material

This section presents a description of the data sources used in this case study of the Greater Washington, DC metropolitan region.

20.3.1.1 The 1999 IMPLAN Database

IMPLAN (IMpact Analysis for PLANning; see IMPLAN 2000) originally was developed by the United States Department of Agriculture (USDA) Forest Service in cooperation with the Federal Emergency Management Agency and the USD of the Interior (USDI) Bureau of Land Management to assist the Forest Service in land and resource management planning. The IMPLAN accounts closely follow the accounting conventions used in the "Input-Output Study of the US Economy" by the Bureau of Economic Analysis (US Department of Commerce; see Guo and Planting 2000) and the rectangular format recommended by the United Nations. The IMPLAN database provides the information needed to create regional input–output tables. It consists of the following two major parts:

1. National-level technology matrices.
2. Estimates of regional data for institutional demand and transfers, value-added, industry output and employment for each county in the USA, as well as state and national totals.

The regional industry-by-industry transactions report is the industry-by-industry analog of the use matrix. It shows, in millions of dollars, the amount purchased from local industries (IMPLAN User's Guide, p. 77). It is generated from the commodity-by-industry formulation of the country's social accounts.

The default trade flow assumptions are Regional Purchase Coefficients (RPCs). RPCs are derived with an econometric equation that predicts local purchases based on a region's characteristics (p. 14). The default is an average RPC assumption that results in RPCs based on an output weighted average for all combined states (p. 41). For any substate region, service sector RPCs are the observed values for the state as constrained by the supply/demand pool ratio. When multiple states are combined, one needs to specify which observed RPCs are used (p. 41). By default, all industries/institutions are treated equally; i.e., each takes an equal proportion of its needs from local sources based on that RPC.

The 1999 IMPLAN Package used includes US totals, Virginia state totals, Maryland state totals, Washington, DC totals, and all related county files.

20.3.1.2 ArcView Business Analyst Datasets (Business Analyst 1999)

The dataset included in the version of Business Analyst used is Dun & Bradstreet Business data. Dun & Bradstreet collects information for more than 10 million private and public US companies. Individual businesses are located by address geocoding, but not all have an exact location. Businesses can be retrieved by their Standard Industrial Classification Code and their location. The Dun & Bradstreet Business data file can be used for locating both competitors and marketing opportunities. These data are current to the second quarter of 1999 update.

The dataset includes the information enumerated in Table 20.5.

The Greater Washington, DC Metropolitan Region

This Washington, DC metropolitan region spans the sates of Maryland (MD), Virginia (VA), and West Virginia (WV), as well as the District of Columbia (DC), and includes the counties from these different states that are enumerated in Table 20.6, and portrayed in Fig. 20.6.

Table 20.5 The dataset for the Greater Washington, DC metropolitan region

– Company Name (actual field name = Name)—The legal, incorporated business name.
– Tradestyle (Trade)—Another name by which a company is known.
Example: ESRI, AT&T, IBM.
– Address (street address)—Physical address of location.
City—City of location.
State—State of location.
ZIP Code (ZIP)—ZIP code of location, includes the four-digit extension to the ZIP code.
Phone—Telephone number with area code.
– Line of business (Linofbus)—Verbal description of the primary SIC code.
– Sales Key Code (Salescod)—Preset ranges for annual sales revenue.
– Status Indicator Code (Statuscd)—whether a location is a single location, headquarter/parent, branch, or division office.
– Employees Here Key Code (Empherecd)—Number of employees at location.
– Franchise Code (Franchcd)—Y or N for franchise indication.
– Accuracy Level Value (ALV)—Accuracy of the geocoding of the location.
M = Matched to Street Address
9 = Matched to ZIP+4 centroid
7 = Matched to ZIP+2 centroid
5 = Matched to 5-digit ZIP Code Centroid
C = Matched to City Centroid
– Key—Internal code used for indexing.

Table 20.6 Composition of the study area[a]

District of Columbia	Prince William County, VA
Calvert County, MD	Spotsylvania County, VA
Charles County, MD	Stafford County, VA
Frederick County, MD	Warren County, VA
Montgomery County, MD	Alexandria City, VA
Prince George's County, MD	Fairfax City, VA
Arlington County, VA	Falls Church city, VA
Clarke County, VA	Fredericksburg city, VA
Culpeper County, VA	Manassas City, VA
Fairfax County, VA	Manassas Park City, VA
Fauquier County, VA	
King George County, VA	Berkeley County, WV[a]
Loudoun County, VA	Jefferson County, WV[a]

[a]The extracted IMPLAN dataset does not include data for these two West Virginia counties. Consequently, the case study in this chapter excludes these two counties from the Washington, DC metropolitan region

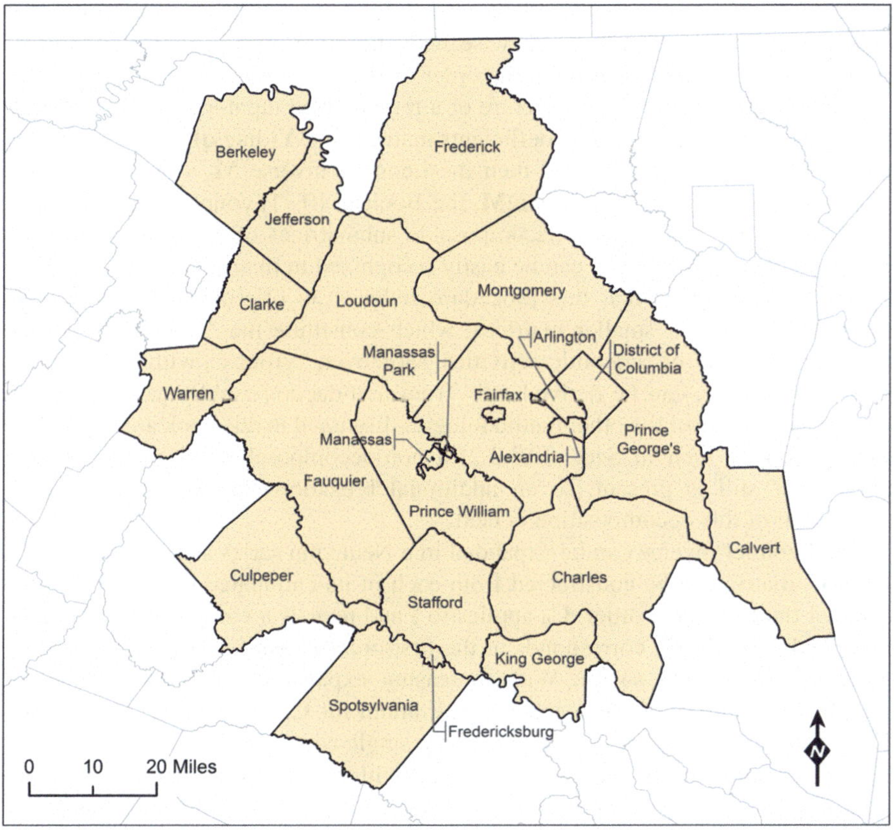

Fig. 20.6 A map of the Greater Washington, DC metropolitan region

20.3.2 Complex Analysis

This section presents the mathematical model first, followed by an application to input–output relations in the Washington, DC metropolitan region.

20.3.2.1 Identifying Complexes

The definition of a complex appears in Sect. 12.1.

Consider a relational (distances, input–output relations) matrix, $\boldsymbol{R}$. Replace all of its values by small numbers, say ε, in order for the Leontief inverse, $(\mathbf{I} - \mathbf{R}^{*})^{-1}$, to exist; $\boldsymbol{R}^{*}$ is the transformed $\boldsymbol{R}$-matrix. Next, count the number of entries, column by column; if the largest number exceeds $n/2$, n being the order of $\boldsymbol{R}$, this constitutes a first complex. If the largest number is smaller than $n/2$, the *initial* group of entries so selected should be inspected for the presence of two, or possibly more, complexes inside; i.e., groups of interdependent entries, but each group being disconnected from the other ones. Continue likewise with the other entries, until the sums of entries of each complex add up to the order of ***R***.

Another way to reveal the structure of a regionalized input–output flow matrix, **Y**, is to compute its associated coefficients matrix, $\mathbf{A} = \mathbf{Y}(\text{diag } \mathbf{q})^{-1}$, where **q** is the vector of production levels, and then the Leontief inverse $\mathbf{M} = (\mathbf{I} - \mathbf{A})^{-1}$. One subsequently constructs **M*** = bool**M**, the Boolean (0−1) counterpart of **M**, and, finally, **C** = **M***′+**M***. The indecomposable submatrices of **Y** are the shown by rows (or columns) of 2s and can be easily recognized in matrix **C**.

One expectation is that this procedure reduces an original ≈400-by-400 **Y**-matrix to a series of smaller matrices, which constitute the "complexes" inside the given "cluster" of economic activities. Moreover, "groups", with their characteristic circularity, can be derived only from non-decomposable matrices, which also is true if some of the other methodologies discussed in this book are applied. In fact, as can be seen in Sect. 20.3.2.2, a non-decomposable sub-matrix of size 371-by-371 still is present, so an additional breakdown appears necessary. A description of this decomposition is next.

The Leontief inverse can be expanded in a Neumann series as $\mathbf{I} + \mathbf{A} + \mathbf{A}^2 + \ldots$, and a $\mathbf{C_i}$-matrix can be constructed from each of its cumulated terms, *i* being the index of the expansion order. $\mathbf{C_0}$ applies to **I** and reveals *n* complexes (in fact, the *n* original sectors); $\mathbf{C_1}$ corresponds to the first-order expansion, $\mathbf{C_2}$ to the second-order expansion, and so on. With increasing expansion order, the number of complexes declines, until it reaches its minimum for $\mathbf{C}_\infty$ (in fact, for the 371-by-371 empirical application, it converges on a single complex).

This technique is applied in Sect. 20.3.2.2 with a first- and second-order expansion. A graph—call it a *complexogram*—could be derived from the number of complexes generated as a function of the expansion order or degree. One expectation is that—like in *p*-median analysis or a scree plot in principal components/factor analysis—somewhere along the expansion axis a downward kink appears in this graph.

20.3.2.2 An Application to the Washington, DC Metropolitan Region

When the preceding technique is applied to the 388-by-388 interactivity flow matrix (the public sector is omitted) for Washington, DC, it reveals a 371-by-371 complex, flanked by *17* 1-complexes; the latter correspond to sectors having either empty columns or a column with a single flow.

If the 371-by-371 multiplier is expanded linearly, it reveals the presence of a much smaller complex (33-by-33), but the quadratic expansion leads to a 369-by-369 one. These results arise because these expansions do not include the totality of feedbacks.

This outcome raises a number of considerations. In fact, not all sectors, alien from a given one, should be considered as potentially "complexifying" with it, as there are no economic or organizational reasons to stick together, locationally speaking. Examples include telephone or electricity bills, the handling of which does not require any close contacts, while, on the contrary, delivery of equipment to these two sectors does. Moreover, very small purchases or deliveries could hardly be considered as complex building, so they too could be removed from the analysis.

In what follows, only the last point is considered, by excluding flows smaller than the minimum of 3% purchases or deliveries with respect to total intermediate purchases or deliveries. These deletions divide the overall multiplier into three parts, $\mathbf{M_1}$, $\mathbf{M_2}$, and $\mathbf{M_{12}}$, corresponding to the important flows, the discarded flows, and the cross-products between these two types of flows.

The following result was quite unexpected: one 111-by-111, one 3-by-3, and three 2-by-2 complex, as well as 251 1-complexes (isolated sectors). This outcome invites a closer study of this situation focusing on whether or not some strange connections exist between the sectors belonging to the large complexes (the 371-by-371 one, and the reduced 111-by-111 one) responsible for the very strong local interconnectedness of the sectors making up the Washington, DC study area economy.

20.4 Chapter Conclusions

If a discipline has worked its way through the analysis of activity complexes, it is for sure employing analytic geometry; certainly, matrix algebra also had to be brought to the fore here, but it is a more classical mathematical framework in the field.

In fact, to get a complete picture of a given complex, both disciplines are required. But given the scope of this chapter, invoking analytical geometry was in order. It allows the disclosing of some hidden characteristics, especially of urban complexes, assembling joint conclusions of spatial statistics and spatial economics, a fine end for this chapter.

References

Beguin, H., & Thisse, J.-F. (1979). An axiomatic approach to geographical space. *Geographical Analysis, 11*, 325–341.

Bunch, B. (1989). *Reality's mirror: Exploring the mathematics of symmetry*. New York: Wiley.

CRC. (1959). *Standard mathematical tables* (12th ed.). Cleveland: CRC Press.

Diestel, R. (2000). *Graph theory* (2nd ed.). New York: Springer.

Fisher, R., & Ziebur, A. (1967). *Integrated algebra and trigonometry, with analytic geometry* (2nd ed.). Englewood Cliffs, NJ: Prentice Hall.

Fylstrom, D., Lasdon, L., Watson, J., & Waren, A. (1998). Design and use of the Microsoft Excel solver. *Interfaces, 28*(5), 29–55.

Getis, A., & Paelinck, J. (2004). On analytical descriptions of geographical patterns. *L'Espace Géographique, 1*, 61–68.

Griffith, D., Paelinck, J., & van Gastel, M. (1998). The Box-Cox transformation: New computation and interpretation. In D. Griffith, C. Amrhein, & J.-M. Huriot (Eds.), *Econometric advances in spatial modelling band methodology. Advanced studies in theoretical and applied econometrics series* (Vol. 35, pp. 45–56). Dordrecht: Kluwer.

Guo, J., & Planting, M. (2000). *Using input-output analysis to measure U.S. economic structural change over a 24 year period* (Working Paper WP2000-01). Washington, DC: Bureau of Economic Analysis, U.S. Department of Commerce.

IMPLAN. (2000). *IMPLAN Pro software*. Huntersville, NC: Implan Group LLC..

Nicolle, J. (1965). *La symétrie*. Paris: Presses Universitaires de France.

Paelinck, J., & Klaassen, L. (1979). *Spatial econometrics*. Farnborough, UK: Saxon House.

Paelinck, J. (1982). Operational spatial analysis: Presidential address. *Papers and Proceedings of the Regional Science Association, 50*, 1–7.

Paelinck, J. (2000). *Controlling complexity in spatial modelling* (Working Paper No. 362). Munich: CESifo Group.

Reinaud, A., van Gastel, M., & Paelinck, J. (1995). Computing Box-Cox transform parameters: A new method and its application to spatial econometrics. In L. Anselin & R. Florax (Eds.), *New dimensions in spatial econometrics* (pp. 136–155). Berlin: Springer.

Rosen, J. (1983). *Symmetry primer for scientists*. New York: Wiley.

Stone, R., & Croft-Murray, G. (1959). *Social accounts and economic models*. London: Bowes and Bowes.

Tobler, W. (1977). Numerical approaches to map projection. In I. Kretschmer (Ed.), *Studies in theoretical cartography* (pp. 51–64). Vienna: Deuticke.

Tutte, W. (2001). *Graph theory*. Cambridge: Cambridge U. Press.

Chapter 21
General Conclusions About Spatial Econometrics

The main experiences reported in this second part of the book can be summarized as follows.

Already mentioned in previous work (Paelinck and Klaassen 1979, pp. 9–11) is that metrics matter. Chapter 12 once more illustrates this statement. The same is true of time, the choice of the correct dynamics being strategic in time-bound spatial econometric modeling, and this is all the more true if policy measures have to be derived via those models.

An important point, which is made in this book when treating one specification of a hybrid dynamical system, is that regional-specific parameters could be—and more often than not are—dependent on the economic state of what may be called a reference region, which most of the time is the total set of regions subjected to an analysis. The relative performances of the latter region are shown to impact the **W** matrix linking elementary regions.

Another aspect of regional diversity is the formation of clusters, for the analysis of which a new concept—connectropy—is introduced in Chap. 16 and then applied. Still another aspect of this nontraditional modeling is the specification of linear expenditure systems. Statistical indicators, binary programming, and complex formation close the rows.

In conclusion, this book again (see Griffith and Paelinck 2011) shows that a lot of nonstandard approaches await a spatial analyst to be tried on ever improving statistical materials, the places where the gold lies are easily reachable, so do not wait to start digging. . . .

Two components of this digging process are worth mentioning in detail.

D. A. Griffith, J. H. P. Paelinck, *Morphisms for Quantitative Spatial Analysis*, Advanced Studies in Theoretical and Applied Econometrics 51,
https://doi.org/10.1007/978-3-319-72553-6_21

21.1 Complexity

Space–time economics—to borrow an expression from theoretical physics—indeed is characterized by great complexity; we refer here not only to the reported study about potentialized partial differential equations (Coutrot et al. 2009) but also to all the other ones, to come to a close with the last exercise.

All of the patterns we inspect here start with theoretical spatial economics, and should flow over into spatial econometrics, if we want our theories to confront the facts, possibly to be contradicted, at times to be relegated to the waiting room of theories pending their ever uncertain status, or to the dust bins of history when they are proven wrong, ...

To get an idea of the relevance of one or another specification, one should look into the complexity of a problem, by which is meant *computational complexity* of the series to be explained (Chaitin 1975; Wolfram 2002, pp. 557–559). This complexity, which we call *conditional complexity*—due to the presence of exogenous variables—can be expressed through the number of parameters necessary to fit a polynomial to an endogenous variable in the exogenous ones. An indicator on [0, 1] is derived in Getis and Paelinck (2004).

One special case of complexity is *non-convexity*, which might be considered constituting a major challenge for spatial econometricians; multiple regimes should be a normal problem to be faced. This book provides an interesting case via research about dynamic micro-regional behavior by means of a partial finite difference equation.

The model has been applied to the most populated region in France after Ile-de-France, the Rhône-Alpes region (Coutrot et al. 2009). In this case, it enables an analysis of the development of knowledge-based industries by means of corresponding employment in activities close to the concept of knowledge-based industries over three time periods in 39 towns of Rhône-Alpes. The results of this analysis appear in Table 21.1.

The results are remarkable from different points of view, but apart from theoretical-econometric considerations, empirical evidence adds robustness to the conclusions, because the two-regimes model class of Table 21.1 includes the three main activity centers of the region, to wit Lyons, Grenoble, and Saint-Etienne, which moreover have a positive constant and, consequently, a positive autonomous acceleration.

Table 21.1 Two regression results

Parameters	A global model	A two-regimes model
A	−0.0041	0.0122
B	−0.0049	0.0072
C	0.0002	−0.0015
a^*	–	−0.0059
b^*	–	−0.0003
c^*	–	0.0008
(Pseudo-)R^2	0.5156	0.9990

Table 21.2 Four-region model for the Netherlands

Spatial units c_t values	NNL	ENL	WNL	SNL
High	0.2127	0.0574	0.0040	0.1205
	0.1120	−0.8495	0.2805	−0.0098
	0.5329	0.0321	−0.5660	0.3123
	−0.8640	0.7897	0.2930	−0.1979
	0.5100	−0.5371	0.1389	−0.0546
Low	−0.7946	−0.0016	0.0791	0.2316
	1.1627	−0.0016	−0.0424	−0.2370
	−0.0515	0.0929	−0.0424	0.0964
	−0.2969	−0.1139	−0.0237	−0.0860
	−0.6304	0.1396	0.3547	−0.0524

21.2 Parameter Relativity

But more astonishing still are the results of the application of an adapted Blondel–Tsitsiklis model (Blondel and Tsitsiklis 1999) as reported in Chap. 14. Recall that the values of the parameters depend on a critical value of the explanatory parameters, in the econometric case treated as a possibly varying quantity.

Estimated as a generalized spatial econometric Lotka–Volterra model (see Griffith and Paelinck 2011, Chap. 10), the model was applied to the four Dutch macro-regions (North, East, West, South: NNL, ENL, WNL, SNL) with c_t being the national growth rate between t and $t + 1$, scaled to $0 - 1$. The objective of this investigation was to analyze how much interregional reactions to regional products depend on overall growth, a weighted average of local growth rates, whereas in the associated inequalities the **c** vector is constant.

Table 21.2 presents, once more, the results with SDLS estimation from Table 14.3 (again Griffith and Paelinck 2011, Chap. 10). Needless to say, the differences between high and low are striking, as well for the orders of magnitude as for the algebraic signs. This result means that it pays to consider the general state of an economy in order to appreciate the interregional links that characterize it. In other words, one has to submit to the *capriciousness* of a system of spatial economic units.

The last two sections clearly show that the real challenge for the future of spatial econometrics is to "create beauty from garbage," as the expression goes . . . .

References

Blondel, V., & Tsitsiklis, J. (1999). Complexity of stability and controllability of elementary hybrid systems. *Automatica, 35*, 479–489.

Chaitin, G. (1975). Randomness and mathematical proof, *Scientific American, 232(5)*, 47–52.

Coutrot, B., Sallez, A., Paelinck, J., & Sutter, R. (2009). On potentialized partial finite difference equations: analyzing the complexity of knowledge-based spatial economic activities. *Région et Développement, 29*, 237–264.
Getis, A., & Paelinck, J. (2004). On analytical descriptions of geographical patterns. *L'Espace Géographique, 1*, 61–68.
Griffith, D., & Paelinck, J. (2011). *Non-standard spatial statistics and spatial econometrics*. Berlin: Springer.
Paelinck, J., & Klaassen, L. (1979). *Spatial econometrics*. Farnborough, UK: Saxon House.
Wolfram, S. (2002). *A new kind of science*. Champaign, IL: Wolfram Media.

Epilogue

Various histories of spatial statistics exist. Griffith (2012a) furnishes one version of it. He emphasizes the major contributions by Cliff and Ord that help establish and popularize this discipline, noting its linkages to spatial econometrics and geostatistics, and discusses current extensions to eigenvector spatial filtering, with special reference to space–time and spatial interaction data. Griffith (2014) also summarizes the diffusion of *spatial statistics* education in the USA, which began slowly in the mid-1980s, but is widely taught throughout that country today. He addresses suitable course content, the interdisciplinary nature of this type of course, and presents a model *spatial statistics* course offering. One outcome of these developments is the emergence of a bi-annual *spatial statistics* conference, first convened in 2013, and the establishment of the journal *Spatial Statistics*. Finally, Griffith (2012b) surveys the interface of spatial statistics and spatial econometrics, with special reference to Griffith and Paelinck (2011). In doing so, he addresses a number of challenges posed by spatial autocorrelation alluded to and/or derived from the spatial statistics topics of their book.

As for spatial econometrics, the years after 1979 have known its progressive practice with principles as advocated in Paelinck and Klaassen (1979) and in many other later publications; the normal incubation period of a new economic discipline is 25 years, as shown by the fact that in 2004, scientist of the University of Zaragoza, Spain (Department of Economic Theory), took the initiative to start a (now) annual seminar about spatial econometrics (for references to that incubation period, one is referred to Sarafoglou and Paelinck (2009). The 10 years after 2004 have shown the fertility of that incubation period with books, seminars, and the founding in 2006 of the Spatial Econometrics Association, which has annual meetings and a Summer School.

This current book contributes to this evolving literature.

D. A. Griffith, J. H. P. Paelinck, *Morphisms for Quantitative Spatial Analysis*, Advanced Studies in Theoretical and Applied Econometrics 51,
https://doi.org/10.1007/978-3-319-72553-6

References

Griffith, D. (2012a). Spatial statistics: A quantitative geographer's perspective. *Spatial Statistics, 1*, 3–15.

Griffith, D. (2012b). Selected challenges from spatial statistics for spatial econometricians. *Comparative Economic Research. Central and Eastern Europe, 15*, 71–85.

Griffith, D. (2014). Reflections on the current state of spatial statistics education in the US: 2014. *Geo-Spatial Information Science, 17*, 229–235.

Griffith, D., & Paelinck, J. (2011). *Non-standard spatial statistics and spatial econometrics*. Berlin: Springer.

Paelinck, J., & Klaassen, L. (1979). *Spatial econometrics*. Farnborough: Saxon House.

Sarafoglou, N., & Paelinck, J. (2009). On diffusion of ideas in the academic world: The case of spatial econometrics (Chap. 4). In T. Nolin (Ed.), *Handbook of regional economics* (pp. 113–128). Hauppage, NY: Nova.

Author Index

D. A. Griffith, J. H. P. Paelinck, *Morphisms for Quantitative Spatial Analysis*, Advanced Studies in Theoretical and Applied Econometrics 51,
https://doi.org/10.1007/978-3-319-72553-6

Subject Index

D. A. Griffith, J. H. P. Paelinck, *Morphisms for Quantitative Spatial Analysis*, Advanced Studies in Theoretical and Applied Econometrics 51,
https://doi.org/10.1007/978-3-319-72553-6